A TRAVELER'S GUIDE TO SPACETIME

AN INTRODUCTION TO THE SPECIAL THEORY OF RELATIVITY

A TRAVELER'S GUIDE TO SPACETIME

AN INTRODUCTION TO THE SPECIAL THEORY OF RELATIVITY

Thomas A. Moore

Pomona College

McGRAW-HILL, INC.

New York St. Louis San Francisco Auckland Bogotá Caracas
Lisbon London Madrid Mexico City Milan Montreal
New Delhi San Juan Singapore Sydney Tokyo Toronto

This book was set in Times Roman by The Clarinda Company.
The editors were Jack Shira and Scott Amerman;
the production supervisor was Leroy A. Young.
The cover was designed by Joseph Gillians.
R. R. Donnelley & Sons Company was printer and binder.

A TRAVELER'S GUIDE TO SPACETIME

An Introduction to the Special Theory of Relativity

This book was printed on recycled, acid-free paper containing 10% postconsumer waste.

1 2 3 4 5 6 7 8 9 0 DOC DOC 9 0 9 8 7 6 5 4

ISBN 0-07-043027-6

Library of Congress Cataloging-in-Publication Data

Moore, Thomas A. (Thomas Andrew)
A traveler's guide to spacetime: an introduction to the special theory of relativity / Thomas A. Moore.
p. cm.
Includes index.
ISBN 0-07-043027-6
1. Special relativity (Physics) I. Title.
QC173.65.M66 1995
530.1'1—dc20 94-39850

ABOUT THE AUTHOR

THOMAS A. MOORE is associate professor of physics at Pomona College in Claremont, California. He received his B.A. magna cum laude with distinction in physics from Carleton College in 1976. He was awarded a Danforth Fellowship in 1976, which supported his graduate education at Yale University. He earned his Ph.D. in physics from Yale in 1981. Professor Moore taught at Carleton College and Luther College before coming to Pomona in 1987. In 1991, Pomona College students voted to give him a Wig Distinguished Teaching Award, the college's highest honor for teaching. In addition to pursuing research in relativistic astrophysics, Professor Moore has been a steering committee member and active participant in the Introductory University Physics Project since 1987, and has served since 1990 as the leader of the development team working on an IUPP model curriculum entitled *Six Ideas That Shaped Physics.* He is currently writing a text for that curriculum. He is married and has two daughters.

For My Parents, Stanley and
Elizabeth Moore, who taught
me the joy of wondering.

CONTENTS

PREFACE

A Traveler's Guide to Spacetime is a textbook on special relativity designed for use as a supplemental text supporting a short section on relativity in virtually any undergraduate college or university physics course: it is appropriate for introductory-level (calculus-based) physics courses as well as sophomore- or junior-level courses in modern physics, classical mechanics, or particle physics. (With some care and judicious omissions, this text might even be used in an algebra-based introductory course.) Preliminary editions have been tested at various colleges and universities in a variety of such courses.

This text is meant to fill a gap in currently available materials on special relativity. Many introductory and modern physics texts have a single chapter on relativity, designed to be "covered" in less than a week of class time. Such chapters are terse by necessity, and thus tend to focus more on laying out the formulas and results of relativity than to helping students understand its logic and meaning. On the other hand, texts devoted to special relativity alone are often too long and/or complicated to use for a brief introduction to relativity as a part of a larger course, and are often written more for the solitary reader than for day-to-day use in the classroom. This text is meant to stand between these extremes.

The experience of those of us who have used preliminary versions of this text is that if two to four weeks of class time are devoted to discussing special relativity using the approach described in this text, students at almost any level can develop a robust understanding of the logic and meaning of relativity, an understanding that is deeper and more satisfying and has a longer half-life than that gained from shorter treatments. Moreover, we have found that students become genuinely excited about being able to understand such a well-known but counterintuitive topic in physics (the intensity of this enthusiasm actually surprised some of our users). A presentation of special relativity with this text thus provides professors with an opportunity to develop students' understanding of this important foundation of modern physics, helps them appreciate the power and beauty of theoretical reasoning in physics, and increases their enthusiasm about physics.

This text has several features that, taken together, distinguish it from available alternatives:

1 *This text is designed for flexible use in the classroom.* Each chapter has been constructed (and tested) to be roughly the right length to serve as the reading assignment for one 50-minute class session. Nearly 150 exercises at a range of levels are provided for use as homework exercises. Material has been carefully arranged between chapters so that certain chapters can be omitted entirely if necessary.

2 *This text develops relativity from a modern, geometrically oriented perspective.* My experience is that the geometric analogy to spacetime and the use of spacetime diagrams really help students understand relativity and remember its results. Unlike most texts, this text uses spacetime diagrams throughout, thoroughly integrating them into the presentation. Students are given explicit and careful instruction in constructing and interpreting such diagrams starting in the second chapter. The text also uses the four-dimensional language and concepts currently used by researchers in astrophysics and particle physics, language that also helps students think clearly and avoid common misconceptions.

3 *This text emphasizes the logical structure of relativity.* The text clearly shows students how well-known relativistic effects such as length contraction and time dilation are the inevitable and logical consequences of the principle of relativity, thus providing a vivid and memorable introduction to the process of theoretical reasoning in physics, as well as some experience with its beauty and power (while also emphasizing the critical importance of experimental verification).

The first chapter begins with an introduction to the principle of relativity ("the laws of physics are the same in all inertial reference frames") and a short discussion of relativity in the context of newtonian mechanics. The second chapter explores the apparent contradiction between this principle and the observed constancy of the speed of light, discusses the concept of clock synchronization, and introduces the spacetime diagram as a tool for visualizing the relationships between events in spacetime.

Misunderstanding of the nature of time and the relativity simultaneity is the single most common source of difficulty for students of relativity. These issues are explored with special care in the text's central chapters (Chapters 3 through 5), emphasizing the metric equation and using the geometric analogy to spacetime to help students anchor these ideas in their own experience. The meaning of the Lorentz transformation is presented in Chapter 6 using spacetime diagrams before *any* equations are developed. Chapters 7 and 8 show how spacetime diagrams can be used to understand length contraction, explain why nothing can go faster than light, and resolve some famous apparent paradoxes.

Chapters 9 and 10 explore the conservation of relativistic momentum and energy, again using diagrams that help students visualize and think about these concepts. Chapter 11 illustrates the application of these principles to particle physics.

POSSIBLE COURSE OUTLINES

At the rate of one chapter per class session, discussing the entire text would take a minimum of 11 sessions (roughly four weeks if there are three sessions per week). But one can construct shorter treatments of the topic by omitting various chapters. A professor desiring the shortest possible introduction to the theory emphasizing *kinematics* might assign the following chapters:

1. Chapter 1: The Principle of Relativity
2. Chapter 2: Synchronization, Units, and Spacetime Diagrams
3. Chapter 3: The Nature of Time
4. Chapter 4: The Metric Equation
5. Chapter 6: Coordinate Transformations
6. Chapter 7: Lorentz Contraction

This gives a reasonably complete look at relativistic kinematics within a two-week time budget. If a few more class days are available, I would recommend first adding Chapter 5, then Chapter 8.

The following plan offers the most compact introduction to relativistic *dynamics:*

1. Chapter 1: The Principle of Relativity
2. Chapter 2: Synchronization, Units, and Spacetime Diagrams
3. Chapter 3: The Nature of Time
4. Chapter 4: The Metric Equation
5. Chapter 6: Coordinate Transformations
6. Sections 5.2 through 5.5 (Proper Time), Section 8.5 (Velocity Transformation)
7. Chapter 9: Four-Momentum
8. Chapter 10: Conservation of Four-Momentum

This scheme provides a solid treatment of relativistic dynamics within a three-week time budget. If more time is available, add Chapter 11 and/or cover more of Chapters 5 and 8.

The basic ideas to keep in mind when constructing the appropriate permutation for your own course are the following:

1. The core material is in Chapters 1 through 4 and Chapter 6: all other chapters draw on the ideas covered in this core.
2. Chapters 9 and 10 should be considered a unit, and have sections 5.2 through 5.5, and section 8.5 (in addition to the core) as prerequisites. Chapters 9 and 10 in turn are prerequisites for Chapter 11.
3. Chapters 7, 8, and 11 are completely optional (except for section 8.5, as noted, which is required for Chapters 9 and 10), and can be included or omitted without affecting continuity. Chapter 5 can also be omitted if Chapters 9 and 10 are: though the material treated there is quite important, it is not *essential* for anything except those chapters.
4. Section 5.10 (on the Doppler shift) and section 8.6 (on the headlight effect) are optional: they are not needed for anything else (although section 11.9 does *refer* to 5.10).

Experience gathered when teaching from the preliminary versions suggests that it is unwise to assign *more* than one chapter per 50-minute class session: each chapter seems to press the limits of what a typical student can absorb in one class session. Indeed, if it is at all possible, it is good to budget an extra day for every three or four chapters to allow for *more* discussion and/or review. The first chapter in particular is rather long, and while much of the material in it is likely to be review for students in some courses, in an introductory course it might be good to spend two class sessions on this chapter (note that there is a logical break between sections 1.7 and 1.8).

Two to four weeks may seem long to spend on a topic that has few applications to the everyday world and that most students who are not physics majors are unlikely to ever "need." To the skeptic, I offer the following comments: (1) My experience is that it really takes about this amount of time for students to understand and appreciate the rather sophisticated arguments involved in relativity. Shorter treatments leave them unsatisfied and their knowledge of the subject insufficiently organized and integrated to survive much longer than the final exam. (2) Students seem to really *enjoy* studying relativity, in spite of its seeming irrelevance. Almost any physics course can benefit from including a topic about which students are really *excited* to learn. (3) Special relativity provides a nearly ideal illustration of the theoretical side of physics: no other topic displays how powerful and beautiful theoretical reasoning can be without also involving difficult, obscuring mathematics. (4) Relativity is one of the greatest triumphs of the human spirit in this century: omitting relativity from a physics course is thus almost like omitting Shakespeare from an English literature course. Why not proudly show our students some of the best that physics has to offer?

ACKNOWLEDGMENTS

Many people have contributed to the production of this text. William Titus at Carleton College provided encouragement and good advice while I was preparing the first versions, and has offered valuable feedback and suggestions for problems as a faithful user of the preliminary editions. Alan Macdonald at Luther College contributed many ideas that have worked their way into the current edition. My students in Physics 101 and Physics 51 at Pomona have been helpful in finding errors and making suggestions, as have Dan Schroeder, Richard Noer, Mara Harrell, Michael Wanke, Doreen Weinberger, Nalini Easwar, Randy Knight, and Woods Halley. I want to thank Kris German for thoughtful criticism and personal support while I was writing various versions and for help with the proofreading. Many thanks to the following McGraw-Hill reviewers: Hendrik Van Dam, University of North Carolina—Chapel Hill; Gordon Feldman, The Johns Hopkins University; Wallace Glab, Texas Tech University; James N. Lloyd, Colgate University; J. Orear, Cornell University; and James H. Smith, University of Illinois at Urbana-Champaign. Thanks also to Peter Stanley, who supported my efforts morally and financially while he was the Dean of the Faculty at Carleton. Many others have contributed to the physical production of various versions, including Debora Gordon, Charlotte Van Ryswk, Tim Lindholm, Janet Runkel, and Connie Wilson. Finally, I'd like to thank my wife Joyce and my daughters Brittany and Allison for their support and understanding during the final push toward publication. Grateful thanks to all!

Thomas A. Moore

1

THE PRINCIPLE OF RELATIVITY

Something old,
Something new . . .

Traditional wedding poem

1.1 INTRODUCTION TO THE PRINCIPLE

If you have ever traveled on a jet airplane, you know that while the plane may be flying through the air at 550 mi/h, things inside the plane cabin behave pretty much as they would if the plane were sitting at the loading dock. A cup dropped from rest in the cabin, for example, will fall straight to the floor (even though the plane moves forward many hundreds of feet with respect to the earth in the time that it takes the cup to reach the floor). A ball thrown up in the air by a child in the seat in front of you falls straight back into the child's lap (instead of being swept back toward you at hundreds of miles per hour). Your watch, the attendants' microwave oven, and the plane's instruments behave just as they would if they were at rest on the ground.

Indeed, imagine yourself being confined to a small, windowless, soundproofed room in the plane during a stretch of exceptionally smooth flying. Is there *any* physical experiment that you could perform entirely within the room (i.e., that would not depend on any information coming from beyond the walls of the room) that would indicate whether or how fast the plane is moving?

The answer to this question appears to be "no." No one has *ever* found a convincing physical experiment that yields a different result in a laboratory moving at a constant velocity than it does when the laboratory is at rest. The designers of the plane's electronic instruments do not have to use different laws of electromagnetism to predict the behavior of those instruments when the plane is in flight than they do when the plane is at rest. Scientists working to enhance the performance of the *Voyager 2* space probe tested out various techniques on an identical model of the probe at rest on earth, confident that if the techniques worked for the earth-based model, they would work for the actual probe, in spite of the fact that the actual probe was moving relative to the earth at 72,000 km/h.[1] Astrophysicists are able to explain and understand the behavior of distant galaxies and quasars using physical laws developed in earth-based laboratories, even though such galaxies and quasars move with respect to the earth at substantial fractions of the speed of light.

In short, all available evidence suggests that we can make the following general statement about the way the universe is constructed:

> *The laws of physics are the same inside a laboratory moving at a constant velocity as they are in a laboratory at rest.*

This is an unpolished statement of what we will call the **principle of relativity.** This simple idea, based on common, everyday experience, is the foundation of Einstein's special theory of relativity. All of that theory's exciting and mind-bending predictions about the nature of space and time follow as *logical consequences* of the principle of relativity! Indeed, the remainder of this book is little more than a step-by-step unfolding of the rich implications of this statement.

The principle of relativity is both a very new and a very old idea. It was not first stated by Einstein (as one might expect) but by Galileo Galilei in a book published in 1632.[2] In the nearly three centuries that passed between Galileo's statement and Einstein's first paper on special relativity in 1905, the principle of relativity as it applied to the laws of *mechanics* was widely understood and used (in fact, it was generally considered to be a *consequence* of the particular characteristics of Newton's laws). What Einstein did was to assert the applicability of the principle of relativity to *all* the laws of physics and most particularly to the laws of electromagnetism (which had just been developed and thus were completely unknown to Galileo). Thus Einstein did not *invent* the principle of relativity; rather, his main contribution was to reinterpret it as being *fundamental* (more fundamental than Newton's laws or even than the ideas about time that up to that point had been considered obvious and inescapable) and to explore insightfully its implications regarding the nature of light, time, and space.

Our task in this text is to work out the rich and unexpected consequences of this principle. But it is important to make two cautionary statements before we proceed:

[1]R. P. Laeser, W. I. McLaughlin, and D. M. Wolff, "Engineering Voyager 2's Encounter with Uranus," *Sci. Am.*, **255** (5), November 1986.

[2]Galileo Galilei, *Dialogue Concerning the Two Chief World Systems—Ptolemaic and Copernican.* A good English translation is the one by Stillman Drake, Berkeley: University of California Press, 1962. In that translation the relevant section starts on p. 186. Galileo's vivid and entertaining statement of the principle of relativity is a wonderful example of a style of scientific discourse that has, unfortunately, become archaic.

(1) The principle of relativity is a *postulate,* and (2) it needs to be more precisely stated before we can extract any of its logical implications.

The principle of relativity is one of those core physical assumptions (like Newton's second law or the law of conservation of energy) that have to be accepted on faith: it cannot be *proved* experimentally or logically derived from more basic ideas (for example, it is not possible even in principle to test *every* physical law in *every* smoothly moving laboratory). The value of such a postulate rests entirely on its ability to explain and illuminate experimental results.

The principle of relativity has weathered intense critical examination for more than 85 years. No contradiction of the principle or its consequences has *ever* been conclusively demonstrated. Moreover, the principle of relativity has a variety of unusual and unexpected implications that have been verified (to an extraordinary degree of accuracy) to occur exactly as predicted. Therefore, while this principle cannot be *proven,* it has not yet been *disproven,* and physicists find it to be something that can be confidently *believed.* The principle of relativity, simple as it is, is a very rich and powerful idea, one that the physics community has found to be not only helpful but crucial in the understanding of much of modern physics.

Turning to the other problem, the principle of relativity as stated earlier suffers from certain problems of both abstraction and ambiguity. For example, what do we *mean* by "the laws of physics are the same"? What exactly do we mean by "a laboratory at rest"? How can we tell if a laboratory is "at rest" or not? If we intend to explore the logical consequences of any idea, it is essential to state the idea in such a way that its meaning is clear and unambiguous.

Our task in the remaining sections of this chapter is to resolve these problems. We will first replace the ambiguous phrases "laboratory," "at rest," and "constant velocity" with a single phrase involving more clearly defined terms. In the final sections, we will explore what we really mean by "the laws of physics are the same" in such laboratories. In so doing, we will provide a firm foundation for exploring the implications of the principle of relativity.

1.2 EVENTS AND SPACETIME COORDINATES

The principle of relativity, as we have stated it so far, asserts that the *laws of physics are the same in a laboratory moving at a constant velocity as they are in a laboratory at rest.* A "laboratory" in this context is presumably a place where one performs experiments that test the laws of physics. How can we more carefully define what we mean by this term?

The most fundamental physical laws describe how particles interact with each other and how they move in response to such interactions. Thus, perhaps the most basic need of a physicist who seeks to specify and test the laws of physics is a means of mathematically describing the *motion* of a particle in space.

In relativity theory we need to be very careful about describing the way that we will measure the motion of particles (hidden assumptions about the measurement process have plagued thinkers both before and after Einstein). In what follows, I will describe how we can measure the motion of a particle in terms of simple and well-defined concepts that are based on a *minimum* of supporting assumptions.

The first of these concepts is described by the technical term *event:*

Definition

An **event** is any physical occurrence that can be considered to happen at a definite place in space and at a definite instant in time.

The explosion of a small firecracker at a particular location in space and at a definite instant in time is a vivid example of an event. The collision of two particles or the decay of a single particle at a certain place and time also defines an event. The simple passage of a particle through a given mathematical point in space can also be treated as if it were an event (simply imagine that the particle sets off a firecracker at that point as it passes by).

Because an event occurs at a specific point in space and at a specific time, we can quantify when and where the event occurs by four numbers: three that specify the location of the event in some three-dimensional spatial coordinate system and one that specifies what time the event occurred. These four numbers are called the **spacetime coordinates** of the event.

Note that the exact values of the spacetime coordinates of an event depend on certain arbitrary choices, such as the origin and orientation of the spatial coordinate axes and what time is considered to be $t = 0$. Once these choices are made and consistently used, however, specifying the coordinates of physical events provides a useful method of mathematically describing motion.

Specifically, we can quantify the motion of a particle by treating it as a series of events. We can visualize this process in the following manner. Imagine an airplane moving along the x axis of some coordinate system. The airplane carries a blinking strobe light. Each blink of the strobe is an event in the sense that we are using the word here: it occurs at a definite place in space and at a definite instant of time. We can describe the motion of the plane by plotting a graph of the position coordinate of each "blink event" vs. the time coordinate of the same, as illustrated in Fig. 1.1.

If the time between blink events is reduced, one gets an even more detailed picture of the plane's motion. By decreasing the time between blinks sufficiently, one can get as detailed a description of the plane's motion as one desires. Therefore, the plane's

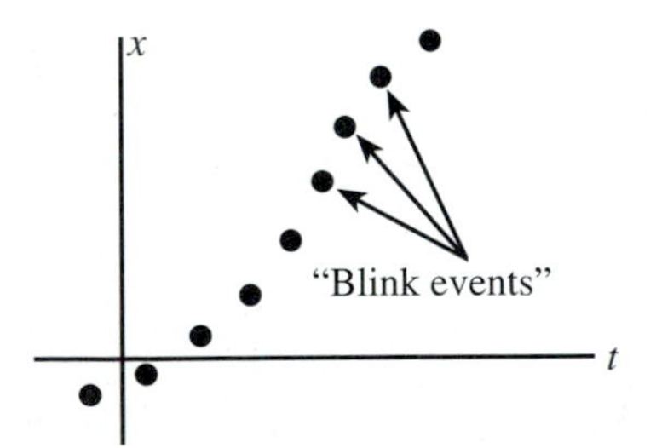

FIGURE 1.1
Sketching out a graph of the motion (position vs. time) of an object by plotting the "blink events" that occur along its path.

motion can be described in arbitrarily fine detail by listing the spacetime coordinates of a sufficient number of blink events distributed along its path.

The above example is a specific illustration of a general idea: the motion of *any* particle can be mathematically described to arbitrary accuracy by specifying the spacetime coordinates of a sufficient number of events suitably distributed along its path. Studying the motion of particles is the most basic way to discover and test the laws of physics. Therefore, *the most fundamental task of a "laboratory"* (as a place in which the laws of physics are to be tested) *is to provide a means of measuring the spacetime coordinates of events.*

1.3 REFERENCE FRAMES IN GENERAL

How might this be done? Imagine constructing a rigid cubical lattice (like a playground jungle gym) consisting of a large number of identical measuring sticks of some given length. At each lattice intersection, you place one of a set of identical clocks, as shown in Fig. 1.2[1]

We will now arbitrarily choose *one* clock in the framework to serve as the **spatial origin** of the reference frame: the origin defines a unique location in the frame against which all other locations are compared. The measuring sticks radiating from the origin

[1]This image of a reference frame as a lattice of clocks has been adapted from E. F. Taylor and J. A. Wheeler, *Spacetime Physics,* San Francisco: Freeman, 1966, pp. 17–19.

FIGURE 1.2
A reference frame visualized as a cubical lattice of clocks and measuring sticks. *(Adapted from E. Taylor and J. A. Wheeler,* Spacetime Physics, *San Francisco: Freeman, 1966, fig. 9.)*

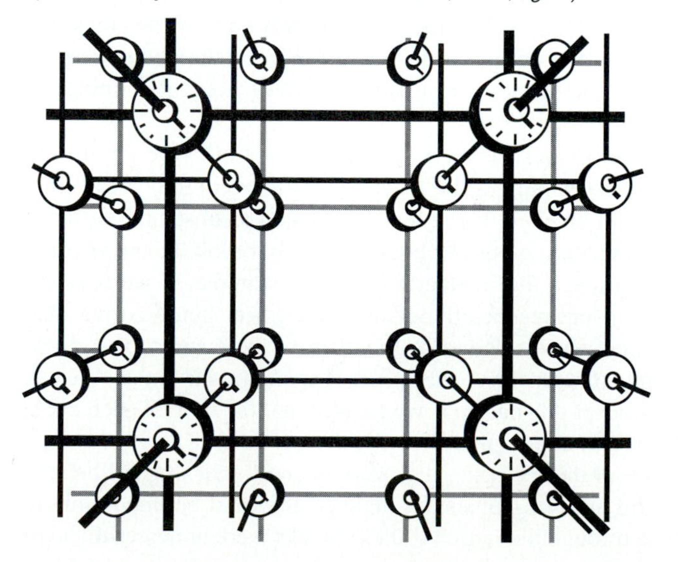

clock specify three mutually perpendicular directions that for the sake of argument we can call the *x direction,* the *y direction,* and the *z direction.* We will take the direction *opposite* to the x direction to be the *negative x direction*, and so on.

Once the origin and these directions have been chosen, the location of any other clock in the reference frame can be uniquely described by stating the distance that one has to travel (as registered by the lattice measuring sticks) along the $\pm x$ direction, the $\pm y$ direction, and then the $\pm z$ direction to get from the origin clock to the second clock. If we agree to always state these numbers in the order $[x, y, z]$, then the location of any clock in the lattice can be specified as *an ordered set of three signed numbers.* For example, the set [-3 m, 6 m, -1 m] specifies the clock 3 m in the $-x$ direction, 6 m in the y direction, and 1 m in the $-z$ direction from the origin. These three numbers are called the **spatial coordinates** of the clock in the reference frame.

The next step is to select an event to represent the **origin of time** $t = 0$. At the instant that this event occurs, all clocks in the lattice should be synchronized (in some manner) to read $t = 0$.

Now imagine that a different event (for example, the explosion of a firecracker) occurs somewhere in the lattice. The *time* of the event is defined to be the time registered by the clock nearest to the event when the event occurs. The *location* of this event in space is defined to be the lattice location of that clock, as specified by its spatial coordinates. (This might not seem to locate the event very precisely if the distance between clocks is large, but we can imagine building a very fine lattice if we need to make very fine measurements.) The measuring sticks in the lattice allow one to directly read the distance of this clock from an arbitrarily chosen lattice origin along the three mutually perpendicular directions defined by the cubical lattice. The **spacetime coordinates** of an event are thus *defined* to be an ordered set of *four* signed numbers: one that specifies the reading of the clock nearest the event and three that specify the spatial coordinates of that clock. One usually specifies the spacetime coordinates of an event with the time coordinate first. We might say that the firecracker explosion occurs at [3 s, -3 m, 6 m, -1 m]; that is, the event is registered by the clock that is 3 m, 6 m, and 1 m away from the origin in the $-x$, $+y$ and $-z$ directions, respectively, and that clock registered the event as occurring 3 s after the event defining the origin of time.

Why is it important to have a clock at every lattice intersection? The point is to make sure that there is a clock essentially *at* the location of any event to be measured. If we attempt to read the time of an event using a clock located a substantial distance away, we need to make additional assumptions about how long it took the information that the event had occurred to reach that distant clock. For example, if we read the time when the *sound* from an event reaches the distant clock, we should correct that reading by subtracting the time it takes sound to travel from the event to the clock. But to do this, we must assume that we know the speed of sound in our lattice. We can avoid making extra assumptions of this nature if we require that the time of each event be measured by a clock that is essentially *present* at the event.

Note that if we must have all these clocks, it is also essential that they all be *synchronized* in some meaningful and self-consistent manner. It would be impossible to track the motion of a particle through the lattice if these clocks were not carefully synchronized, as adjacent clocks might differ wildly and thus give a totally incoherent picture of when the particle passes various lattice points. The appropriate method for synchronizing lattice clocks is actually a subtle issue that we will explore in more detail later. For now, it is sufficient to recognize that it *must* be done.

Once we have specified a synchronization method, the clock-lattice image just described represents a complete definition of a *procedure* that one can use, in principle, to determine the spacetime coordinates of an event. This amounts to what is called an **operational definition** of these spacetime coordinates. In general, an operational definition of a physical quantity defines the quantity by describing how the quantity may be *measured.* Operational definitions provide a useful way of anchoring slippery human words to physical reality by linking the words to specific, repeatable procedures rather than to vague comparisons or analogies.

The procedure just described represents an admittedly idealized method for determining the spacetime coordinates of an event. The actual methods employed by physicists may well differ from this description, but these methods should be *equivalent* to what is described above: the clock-lattice method defines a standard against which actual methods are to be compared. It is such a simple and direct method that it is inconceivable that any actual technique could yield different results and still be considered correct and meaningful.

But the real importance of the clock-lattice definition of spacetime coordinates is as a mental image that tersely and cleanly describes *exactly* what has to be done to determine the spacetime coordinates of an event (and thus quantify particle motions) without the obscuring complications that always arise in building real devices. As a simple and vivid description of the bare necessities required to determine event coordinates, this mental image will make arguments that follow more straightforward.

With this in mind, we define the following technical words to aid us in future discussions:

Definition

A **reference frame** is defined to be a rigid cubical lattice of clocks as described above *or its equivalent,* with some self-consistent method of synchronizing clocks specified.

Definition

The **spacetime coordinates** of an event in a given reference frame are defined to be an ordered set of four numbers: one specifying the time of the event as registered by the nearest clock in the reference frame clock lattice and three specifying the spatial coordinates of that clock in the lattice.

Definition

An **observer** is defined to be a (possibly hypothetical) person who interprets the measurements made in a reference frame (i.e., the person who interprets the spacetime coordinates collected by a central computer from all the clocks).

A reference frame is often spoken of in connection with some object. For example, one might refer to "the reference frame of the surface of the earth" or "the reference frame of the cabin of the plane" or "the reference frame of the particle." In these cases, we are being asked to imagine a clock lattice (or equivalent) fixed to the object in question.

Sometimes the actual reference frame is referred to only obliquely, as in the phrase "an observer in the plane cabin finds . . . ." Since the word *observer* in this text refers to someone who is using a reference frame to make measurements of the coordinates of events, the existence of a reference frame attached to the cabin of the plane is presumed.

A reference frame may be moving or at rest, accelerating, or even rotating about some axis. The beauty of the definition of spacetime coordinates given above is that measurements of the coordinates of events (and thus measurements of the motion of objects) can be carried out in a reference frame no matter how it is moving, provided that the clocks in the frame can be reasonably synchronized in some manner.

Note that the act of "observing" in the context of the last definition is an act of *interpretation* of measurements generated by the reference frame apparatus and may have little or nothing to do with what that observer *sees* with his or her own eyes. When we say that "an observer in such-and-such reference frame observes such-and-such," we are actually referring to the *conclusions* that the observer draws from the coordinate measurements generated by that reference frame lattice.

1.4 INERTIAL REFERENCE FRAMES

Replacing the vaguely defined concept of a "laboratory" in the statement of the principle of relativity with the precisely defined concept of a "reference frame" would represent a substantial improvement in the clarity of that statement. We now turn to the problem of clarifying the ambiguities in the concepts of laboratories "at rest" and laboratories moving at a "constant velocity." How can we operationally define when a laboratory is physically at rest?

The answer is that we cannot! The principle of relativity specifically states that laboratories moving with a constant velocity are physically equivalent to a laboratory at rest. Therefore, there can be no physical basis for distinguishing a laboratory at rest from another moving at a constant velocity. Imagine that you and I are in spaceships coasting at a constant velocity in deep space. You will consider yourself to be at rest, while I am moving by you at a constant velocity. I, on the other hand, will consider myself to be at rest, while *you* are moving by *me* at a constant velocity. According to the principle of relativity, there is no physical experiment that can resolve our argument about who is at rest. Which of us we consider to be at rest is arbitrary.

On the other hand, there are definitely laboratories where the laws of physics are clearly *not* the same as in other laboratories. Imagine yourself to be in a jet accelerating for takeoff. You seem to experience a magical force pressing you into the back of your chair. Unlike real physical forces (such as electric or magnetic forces), this force does not appear to express an interaction between you and any other object in your vicinity; it seems simply to magically appear and disappear. If you were to experience such a force while sitting at home reading a book, you would be deeply disturbed! Yet you are *not* disturbed by this in the plane because you know from experience that the usual laws of physics are violated in an *accelerating* airplane.

The real issue is thus not whether a laboratory is moving or at rest, but *how* that laboratory is moving. In the remainder of this section, we will develop an operational means of determining whether the standard laws of physics hold in a given laboratory or not without using the terms *rest* or *constant velocity,* indeed without referring to anything *outside* the laboratory at all.

Newton's first law of motion may be stated as follows:

> *An object that experiences zero external force will move in a straight line at a constant velocity.*

Specifically this law implies that if it is known that there are *no* physical forces on an object and it is *at rest* at a certain time, it will *remain at rest* for all time.

With this in mind, we can imagine constructing a device that can test the validity of Newton's first law. Imagine a ball floating in a vacuum in the center of a spherical shell. Imagine that the device has some mechanism that we can use to hold the ball at the center of the shell and then release it at rest so that it floats in the center of the shell. If the ball drifts away from the center and contacts the touch-sensitive surface of the shell, the mechanism can be reactivated to reset the ball at rest in the center (Fig. 1.3).

If the outer shell of this device is a good electrical conductor, the space inside will be shielded from any external electromagnetic fields. If the ball floats in a vacuum and maintains a distance of more than a few micrometers from the outer shell, it will not be affected by forces due to air pressure, sound waves, contact forces, external nuclear forces (which have a very short range), and so on. If we operate this device in deep space, far from any massive objects, it will also not be affected by gravitational forces.

If this is so, the ball should then be completely isolated from *any* external forces. Newton's first law then predicts that if the ball is released at rest in the center of its spherical shell, it will remain exactly at rest at the center of the shell. If Newton's first law is not true, then the ball will eventually drift away from the center and touch the

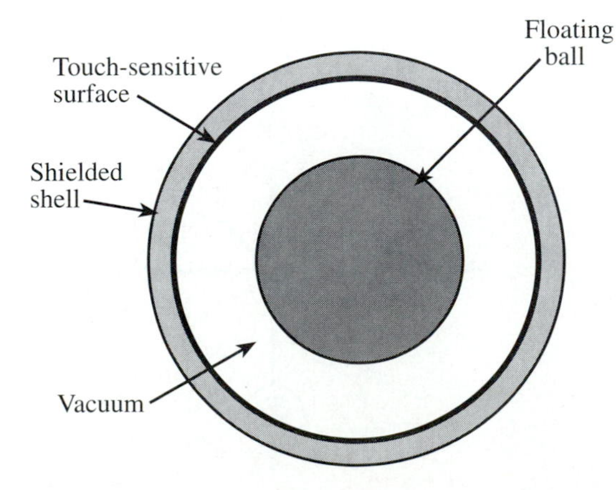

FIGURE 1.3
Cross section of a hypothetical first-law detector. A ball floats in the center of a touch-sensitive spherical shell which shields it from outside forces. If the ball drifts away from the center and touches the outer shell, a mechanism (not shown) is activated that resets the ball at the center of the shell and then releases it at rest.

sensitive inner surface of the detector shell and be reset to the center. The reset action would thus signal a violation of Newton's first law. The frequency of these reset actions will indicate the degree to which that law is violated.

This is just one example of how a first-law detector might (in principle) be constructed; you may be able to think of other approaches that could be used. But if we assume that such a detector can be built, we can make the following definitions.

Definition

An **inertial clock** is a clock moving in such a manner that a first-law detector constructed as described above (or its equivalent) fixed to the clock detects no violation of Newton's first law.

Definition

An **inertial reference frame** is a reference frame whose lattice clocks are all inertial clocks. Equivalently, an inertial reference frame is a reference frame in which Newton's first law is measured to be true at *every* point in the frame.

Again, note that the definition of an *inertial reference frame* amounts to being an *operational definition,* because we have defined *reference frame* by describing *how* to build one (see the definition of reference frame in Sec. 1.3) and *inertial* by describing explicitly *how* one might test a given frame to determine whether it qualifies as being inertial.

The definition of an inertial reference frame given above makes it easy to distinguish inertial and noninertial frames in realistic circumstances. For example, a frame attached to an airplane accelerating for takeoff obviously *cannot* be an inertial frame; even without actually trying it, you know that the ball in the center of a first-law detector in such an accelerating plane will spontaneously drift toward the rear of the plane. Similarly, a reference frame floating in deep space but rotating about some axis cannot be an inertial frame; you know that the ball in a first-law detector will drift away from the center of rotation in such a frame (see Fig. 1.4).

FIGURE 1.4
First-law detectors in *(a)* a linearly accelerating reference frame and *(b)* a rotating reference frame. The frames are represented schematically by a set of coordinate axes. The drift directions of the floating balls inside the first-law detectors are also represented schematically.

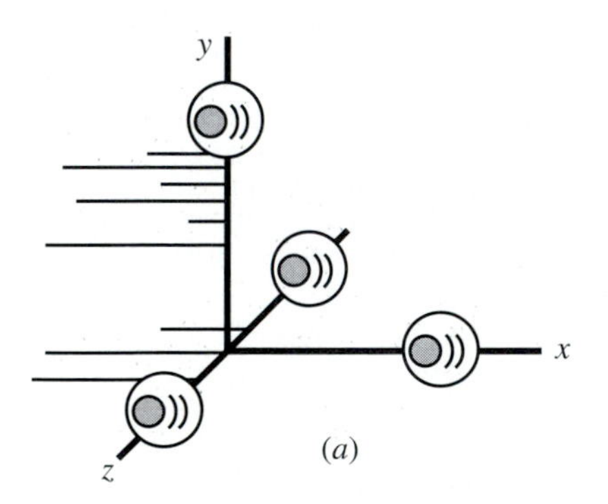

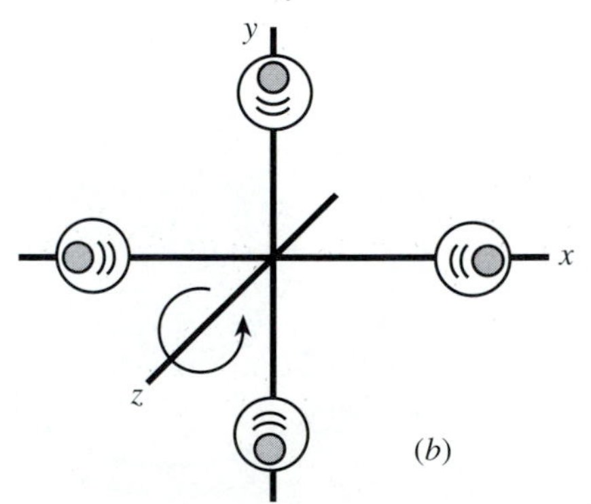

A good example of an inertial reference frame would be a reference frame floating freely in deep space. You can easily imagine that if you placed first-law detectors at various places in such a frame and reset each detector ball to rest in the center of its shell, each ball would continue to float at rest, fulfilling the terms of the definition of an inertial reference frame.

1.5 THE RELATIVE VELOCITY BETWEEN TWO INERTIAL FRAMES

The following statement is an immediate consequence of the definition of an inertial reference frame given in the previous section (as we will see in a moment):

> Any inertial reference frame will be observed to move with a *constant velocity* by observers in any other inertial reference frame. Conversely, a rigid reference frame that moves at a constant velocity with respect to any inertial frame must itself be inertial.

This statement is very important! The fact that *inertial frames move at constant velocities with respect to each other* will be used over and over as we work out the consequences of the principle of relativity. This statement also provides an easy way to distinguish between inertial and noninertial frames: the second part of the statement tells us that if we happen to know that any specific reference frame is inertial, *then any frame that moves at a constant velocity with respect to that inertial frame must also be inertial.* For example, if a frame attached to the ground *is* inertial, then a frame attached to a plane accelerating for takeoff is *not* inertial because it is accelerating horizontally relative to the ground, while a frame attached to a plane flying horizontally at a constant velocity *is* inertial.

Here is an argument for the *first* part of the statement (the proof of the converse statement is left as an exercise in Prob. 1.3). Imagine two inertial reference frames, which we will call the *Home Frame* and the *Other Frame* (see Fig. 1.5). "Home Frame" and "Other Frame" are capitalized in this text to emphasize that these phrases are actually *names* of inertial frames. Since these frames are inertial, Newton's first

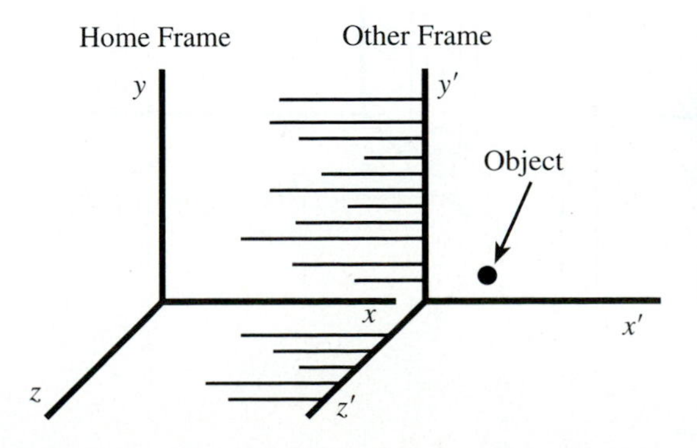

FIGURE 1.5
A force-free object at rest in the Other Frame must move at a constant velocity with respect to the Home Frame by Newton's first law. But since the object is at rest with respect to the Other Frame, that frame itself will be observed to move with the *same* constant velocity with respect to the Home Frame.

law is true in both by definition. Consider an object with no external forces acting on it that also happens to be at rest in the Other Frame. By Newton's first law, it will *remain* at rest in that frame. Now observe the same object from the Home Frame. Since the object has no forces acting on it, Newton's first law implies that it must move at a *constant velocity* in that frame. But if that object is at rest in the Other Frame, the whole Other Frame will be observed in the Home Frame to move with the *same* velocity as the object! Therefore, *the Other Frame will be observed to move with a constant velocity with respect to the Home Frame,* as stated above.

1.6 REFERENCE FRAMES IN THE PRESENCE OF GRAVITY

Reference frames in a gravitational field, however, present a special problem. Consider a frame at rest on the earth's surface. Such a frame *fails* the first-law test because the ball in the center of a first-law detector will *not* float at rest but instead drift (i.e., fall!) toward the earth (see Fig. 1.6*a*).

On the other hand, a reference frame that is itself freely falling *is* inertial by this definition. Consider a nonrotating reference frame that is near the earth but freely falling toward it or around it (e.g., a reference frame inside a freely falling elevator or inside an orbiting space shuttle). In such a frame all objects appear to be weightless. If you release such an object from rest, it will remain at rest, floating in the air. If you

FIGURE 1.6
If we apply the definition of inertial frame literally, a reference frame *(a)* at rest on the earth's surface is *not* an inertial reference frame, since the ball in a first-law detector will not float at rest in the center of its shell. On the other hand, a freely falling reference frame *(b) is* an inertial frame: the ball in a first-law detector will float in the center of its shell. (An observer on the ground would say that this is because the ball and frame fall at exactly the same rate.)

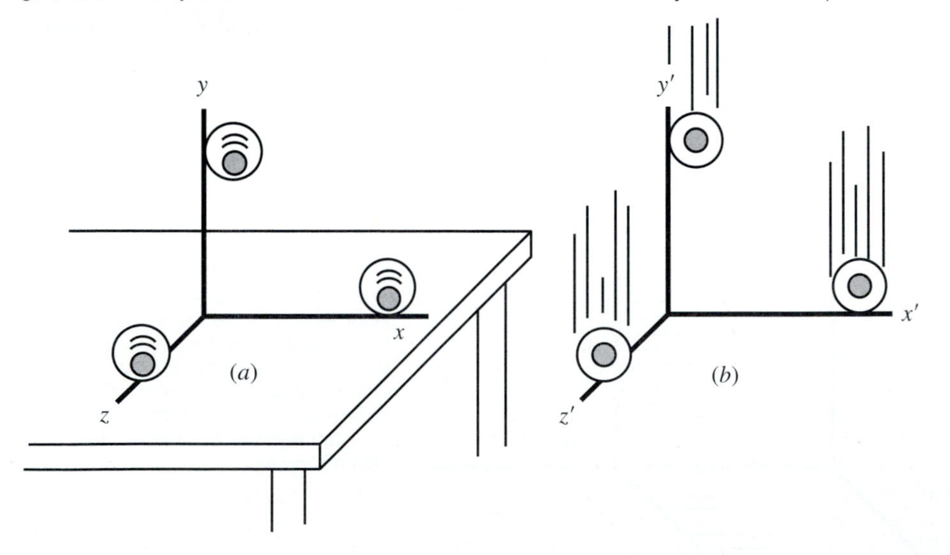

give it a brief push, it will move off afterward in a straight line at a constant velocity. You can easily imagine that if you set up a first-law detector in such a frame, the ball in such a frame will float nicely in the center of its shell (see Fig. 1.6*b*).

Of course we might argue that the drift in the first case is really not a result of a violation of Newton's first law but instead reflects the presence of an external gravitational force acting on the detector ball. In the case of the freely falling frame, the reason that the ball seems to float is because ball, detector, and frame all fall with the same acceleration. If we accept this reasoning, then we have another problem. Since we *cannot* shield the mass from the effects of gravity, the first-law detector does not really work in a gravitational field, and therefore we have no way of defining an inertial reference frame in the presence of a uniform gravitational field!

It turns out that there are sound physical reasons for resolving this problem in a simple and yet radical manner. Ignore the presence of this assumed gravitational "force" and simply apply the definition of an inertial frame *literally.* If a first-law detector in a frame registers drift, then the frame is *not* inertial; if it does not register drift, then the frame *is* inertial. According to this definition, a frame at rest on the earth's surface is *not* strictly inertial. In 1907, Einstein published a paper on gravitation in which he argued that a reference frame at rest in a (hypothetical) *uniform* gravitational field should be physically equivalent to a uniformly *accelerated* (i.e., *non*inertial) reference frame in deep space, and proposed experiments that would test this claim. The results from such experiments support Einstein's assertion that freely falling frames are inertial and frames at rest on the earth's surface are noninertial. Though it would go too far afield from the main line of the argument at this point to discuss these experiments in detail, suffice it to say that even in the presence of a gravitational field, applying our definition of inertial frame literally makes good physical sense.

The only problem is that any *real* gravitational field has nonuniformities that mean that a falling reference frame is only *locally* inertial: first-law detectors near the *center* of a falling reference frame register no violation of Newton's first law, but the farther away a detector is from the center, the more the detector ball will be seen to drift away from rest. To see why this is, consider a detector at the center of mass of the falling frame. Since both the detector's ball and frame fall with the same acceleration, the ball will remain at rest with respect to the frame, and the detector will register no violation of Newton's first law. Now consider a detector located a significant distance above the frame's center of mass. Since the detector ball in this case is somewhat farther from the center of the earth than the frame's center of mass is, it will fall very slightly slower than the frame as a whole and thus will tend to drift upward with respect to the detector shell.

Even so, a nonrotating, freely falling frame is the best approximation to an inertial frame that can be found in the vicinity of the earth. Unfortunately (or perhaps fortunately, depending on how you look at it), very few physics laboratories are freely falling: most are at rest on the surface of the earth. How does the principle of relativity relate to such ordinary laboratories? We will see in a later chapter that the most important consequences of the principle of relativity arise from arguments concerning the synchronization of clocks that are not affected much by the presence of gravity. So in terms of these effects, even a frame in a weak gravitational field (e.g., the field near the earth as opposed to near a black hole) can be considered to be *approximately* inertial.

In any case, the gravitational field acts in the vertical direction alone; if we focus our attention on reference frames and objects that move only horizontally, then gravity has no important effect at all. We might imagine modifying our first-law detectors by replacing the floating ball with an air puck floating on a cushion of air above a horizontal surface. The drift of this air puck would detect any violations of Newton's first law in the horizontal plane and could be used to distinguish between inertial and noninertial frames moving horizontally on the earth's surface.

To summarize, here is a list of the three types of inertial frames that we have discussed, in decreasing order of "perfection":

1. The ideal inertial reference frame is a nonrotating frame that floats freely in space infinitely far from any gravitating objects. Such a frame is *perfectly* inertial and should be your mental image of an inertial frame in the remainder of the text. Such frames are entirely inaccessible in practice.
2. Nonrotating and freely falling frames are *locally* inertial and represent the best approximation to inertial frames that are available in practice. The larger such a frame is, the more noninertial it will be at its periphery. The cabin of the space shuttle is a reasonably accurate inertial frame.
3. Frames at rest on the earth's surface are less accurately inertial but are good enough for most practical purposes, especially if one is only considering motion in the horizontal plane. A modified first-law detector consisting of an air puck floating above a horizontal plane might be used to check for violations of Newton's first law in that plane.

1.7 THE FINAL VERSION OF THE PRINCIPLE OF RELATIVITY

Notice that we have *defined* inertial reference frames so that at least one physical law (Newton's first law) is true in every such frame. The concept of an "inertial reference frame" also represents a precisely and operationally defined replacement for problematic "laboratories at rest" and "laboratories moving with a constant velocity." The concept of "reference frame" precisely and efficiently captures the sense of a "laboratory" as a place where one does physical experiments.

This prompts us to rephrase the principle of relativity as follows:

The Principle of Relativity

The laws of physics are the same in every inertial reference frame.

This is our final polished statement of the principle of relativity. It replaces the fuzzy, ambiguous concept of a "smoothly moving laboratory" with the well-defined concept of an "inertial reference frame." What we are claiming is that if Newton's first law is obeyed in the set of reference frames that are "inertial," then *all* the laws of physics will be obeyed in that set of frames.

But what exactly do we mean when we say that "the laws of physics are the same" in all inertial frames? The purpose of the last three sections of this chapter is to explore what this statement means in a bit more detail. In the next two sections, we will make an *assumption* about the nature of time that will allow us to compare quantities (like the velocity or acceleration of an object) in two different inertial frames. This will allow us in Sec. 1.10 to explore exactly how the laws of ordinary newtonian mechanics can be "the same" in different inertial frames.

1.8 THE NEWTONIAN SOLUTION TO CLOCK SYNCHRONIZATION

An inertial frame floats in space, ready to use. We would like to use it to measure coordinates of events happening within it. But an important problem remains to be solved. How do we synchronize all those clocks? If the clocks are not synchronized, the time coordinate of an event will be meaningless: it will depend in an arbitrary, random way on which actual clock in the lattice records the event.

"The solution is easy," says a newtonian physicist. "Everyone knows that time is absolute and flows equably without regard to anything external. Any good clock will therefore measure the flow of this absolute time. Therefore, simply designate one clock to be a master clock, carry it around to each of the lattice clocks, and synchronize each lattice clock to the master. Since the master clock and the lattice clocks all measure the flow of immutable absolute time, the motion of the master clock as it is carried from place in the lattice is irrelevant. Once a lattice clock is set to agree with the master clock, it will certainly remain in agreement with it, since both clocks measure the flow of absolute time. Indeed, if the master clocks in two *different* reference frames are in agreement at any given event, then *all* the clocks in the two frames will *always* agree. It does not matter whether the frames are in motion with respect to each other; it does not even matter if they are inertial or not. This follows from the self-evident absolute nature of time."

This picture of the nature of time is straightforward and believable. It reflects the intuitive picture of time that most of us already hold. But what are its consequences?

1.9 GALILEAN RELATIVITY

Again consider two inertial reference frames called the Home Frame and the Other Frame. We will often (but not always) imagine ourselves in the Home Frame (so that this frame appears to *us* to be at rest). The Other Frame must be moving at a constant velocity $\vec{\beta}$ with respect to the Home Frame, according to the proof given in Sec. 1.5. [The Greek letter β *(beta)* will be used in this text consistently to represent the speed of one frame with respect to another, standing for the "boost" in speed needed to go from being at rest in one frame to being at rest in the other.]

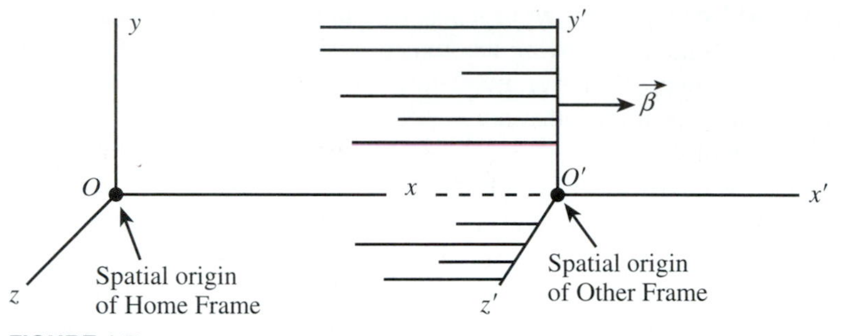

FIGURE 1.7
A schematic drawing of two reference frames in standard orientation. The spatial origins of the frames coincided at $t = t' = 0$ just a little while ago. (You should imagine the frame lattices intermeshing so that events can be recorded in both frames.)

These frames might in principle have any relative orientation, but it is conventional to use our freedom to choose the orientations to put the two frames in **standard orientation,** where the Home Frame's x, y, and z axes point in the same directions as the corresponding axes in the Other Frame. The Home Frame and Other Frame axes are distinguished by referring to the Home Frame axes as x, y, z and the Other Frame axes as x', y', and z' (the mark is called a *prime*). It also is conventional to define the origin event (the event that defines $t = 0$ in both frames) to be the instant that the spatial origin of one frame passes the origin of the other. Finally, the common x axis is conventionally chosen so that the Other Frame moves in the $+x$ direction with respect to the Home Frame, implying that the Home Frame moves in the $-x$ direction with respect to the Other Frame (signs in many equations in this text depend on this convention, so it is important to be aware of it and use it consistently). Standard orientation is illustrated in Fig. 1.7.

Now consider an object moving in space that periodically emits blinks of light. Let the spatial position of a certain blink event be the vector $\vec{r}$ as measured in the Home Frame and $\vec{r}'$ as measured in the Other Frame. Since (according to our assumption) time is universal and absolute, observers in both frames should agree at what time this blink event occurs: $t = t'$. The position of the spatial origin of the Other Frame in the Home Frame at that time is simply $\vec{\beta}t$, since the Other Frame moves at a constant velocity $\vec{\beta}$ with respect to the Home Frame, and we conventionally take both frame's origins to coincide at $t = 0$. The relationship between the object's position vectors in the two frames at the time of the blink is (as shown in Fig. 1.8) given by $\vec{r} = \vec{r}' + \vec{\beta}t$, or

$$\vec{r}' = \vec{r} - \vec{\beta}t \tag{1.1}$$

For frames in standard orientation, $\vec{\beta}$ points in the $+x$ direction and Eq. (1.1) can be written

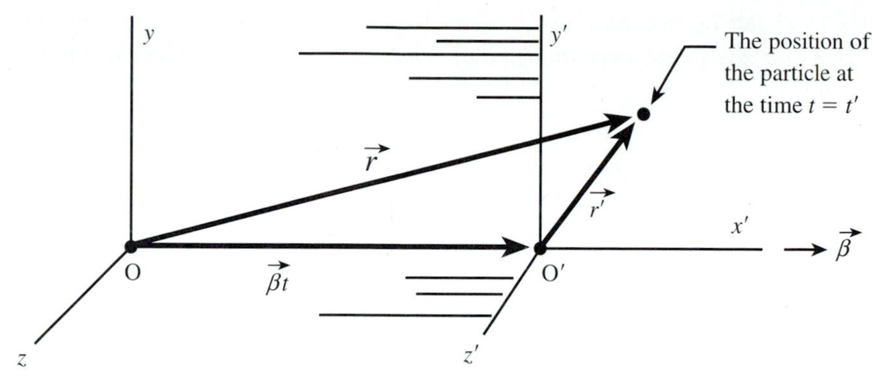

FIGURE 1.8
The relationship between $\vec{r}$ and $\vec{r}'$ for two inertial frames.

$t' = t$	a remainder that time is universal and absolute	(1.2*a*)
$x' = x - \beta t$	*x* component of Eq. (1.1)	(1.2*b*)
$y' = y$	*y* component of Eq. (1.1)	(1.2*c*)
$z' = z$	*z* component of Eq. (1.1)	(1.2*d*)

These equations are called the **galilean transformation equations** and express the consequences of the assumption that time is universal and absolute. These equations allow us to find the position of the object as a function of time in the Other Frame if we know its position as a function of time in the Home Frame.

As time passes, the position of the object will change in time. Thus the position vectors $\vec{r}$ and $\vec{r}'$, though described above in the context of a single blink event along the object's path, are in fact functions of time: $\vec{r}(t)$ and $\vec{r}'(t)$ (since $t = t'$, it does not matter whose time we use to describe these functions). This means that the position coordinates on both sides of Eq. (1.2*b*) to (1.2*d*) are functions of time. If we take the derivative of these equations with respect to this universal and frame-independent time, we get

$$v'_x = v_x - \beta \tag{1.3a}$$
$$v'_y = v_y \tag{1.3b}$$
$$v'_z = v_z \tag{1.3c}$$

which tell us how to find the *velocity* of an object in the Other Frame given its velocity in the Home Frame: these equations are called the **galilean velocity transformation** equations. If we take the time derivative of Eq. (1.3), we get

$$a'_x = a_x \tag{1.4a}$$
$$a'_y = a_y \tag{1.4b}$$
$$a'_z = a_z \tag{1.4c}$$

Equations (1.4) tell us that *observers in both frames will agree about an object's acceleration at a given time,* even though they will disagree about its velocity or position at that time!

1.10 NEWTONIAN RELATIVITY

These equations allow us now to more fully discuss what is meant by "the laws of physics are the same in every inertial reference frame," at least in the context of the newtonian assumption of universal and absolute time.

Consider the following example. A child on an airplane throws a ball vertically into the air and catches it again. As measured in the plane cabin (which we will take to be the Other Frame), the ball appears to travel vertically along the y axis. Now imagine that you watch this process from a nearby mountaintop as the plane passes by. Instead of observing the ball travel vertically up and down, you will instead observe the ball to follow a shallow parabolic trajectory, because in your frame (which we will take to be the Home Frame) the ball, plane, and child all have a considerable horizontal velocity (see Fig. 1.9).

The motion of the ball in these two reference frames looks very different: it looks entirely vertical in the plane's cabin but parabolic in your frame. Even so, you and an observer on the plane would agree (1) that the ball has a certain mass m (which you

FIGURE 1.9
(a) A child throws a ball vertically upward and downward in the cabin of an airplane flying at a constant horizontal velocity. *(b)* From your vantage point on a nearby mountaintop, the ball seems to you to follow a shallow parabolic trajectory because of the ball's large initial horizontal velocity (which is the same as the plane's).

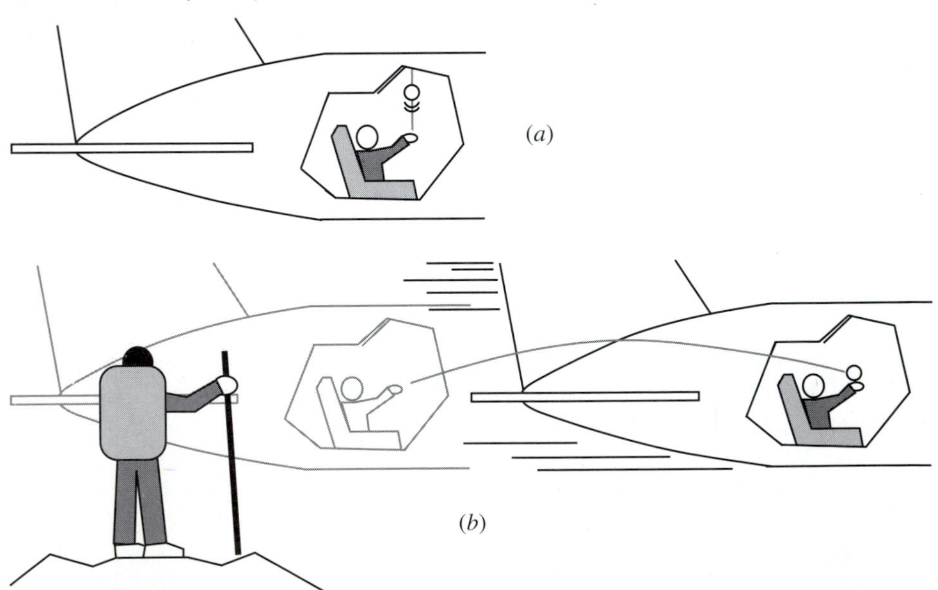

and the observer could each measure with your own balances) and thus should have a gravitational force of magnitude mg acting on it, (2) that this force must be the *net* force on the object while it is in flight (since nothing else is in contact with the ball, ignoring air friction), and (3) that the ball has the same *acceleration* in your respective reference frames [see Eq. (1.4)]. Since you agree on the value of m, the magnitude and direction of the net force on the ball, and the acceleration that the ball experiences, you will agree that Newton's second law

$$\vec{F}_{\text{net}} = m\vec{a} \tag{1.5}$$

accurately predicts the ball's motion, even if you disagree about the initial velocity of the ball and thus the exact character of its subsequent motion (i.e., whether it is vertical or parabolic).

In a similar fashion, you might imagine observing a game of pool in the plane cabin. You on the mountaintop and your friend on the plane will totally disagree about the initial and final velocities of the balls in any given collision (since you will observe them to have a large horizontal component of velocity that your friend does not observe). Even so, you both will find that the total momentum of the balls before a given collision is equal to the total momentum of the balls afterward, consistent with the law of conservation of momentum (see Probs. 1.14 and 1.15)

This is what it means to say that the "laws of physics are the same" in different inertial frames. Observers in different inertial frames may *disagree* about the values of various quantities (particularly velocities), but each observer will *agree* that if one takes the mathematical equation describing a physical law (like Newton's second law) and plugs in the values measured in that observer's frame, one will always find that the equation is satisfied. In other words, *the same basic equations will be found to describe the laws of physics in all inertial reference frames.*

1.11 SUMMARY

Let us review the main issues presented in this chapter.

1 We began by discussing the principle of relativity, stating it in intuitive language and presenting a couple of examples.
2 We then set out to refine this statement by more carefully describing what we mean by a "laboratory." The basic task of a laboratory is to measure the motion of objects. This can be accomplished using a *reference frame,* which uses a cubical lattice of clocks and measuring sticks (or its equivalent) to quantify the position and time of *events,* occurrences (such as blinks of an airplane wing light or firecracker explosions) that happen at a well-defined point in space and instant in time. The set of four signed numbers that describe the time and position of an event are called the *spacetime coordinates* of that event in the frame in question. The motion of an object can then be quantified by treating it as a sequence of events. An *observer* is someone who collects and interprets information about events as measured in a given frame.

3 A reference frame is said to be *inertial* if first-law detectors placed throughout the frame all register no violation of Newton's first law. Such inertial frames move at constant velocities with respect to each other. The precisely defined concept of an "inertial reference frame" replaces the ambiguous idea of laboratories "at rest" and "moving at a constant velocity." The principle of relativity can thus be precisely stated as follows: *The laws of physics are the same in every inertial reference frame.*

4 We then saw that *if* we assume that time is "universal and absolute," it is possible to link the spacetime coordinates of an event in one frame to those of the same event in a different frame using the *galilean transformation equations.* Using these equations, we found in a specific example that observers in different inertial frames would agree that Newton's second law holds in both frames. This concretely illustrates the meaning of principle of relativity.

The only problem with all of this is that the assumption that time is "universal and absolute" turns out to be *wrong*! We will discuss the evidence for this incredible assertion and some of its implications in the next chapter.

PROBLEMS

1.1 DETECTING MOTION. One can usually determine whether an airplane is in flight or not by determining whether the plane is jostling: if the plane is jostling, it is probably either rolling on rough ground or flying through turbulent air. If it is not, it is *likely* to be at rest on the earth. When the plane is jostling, should it be considered a "smoothly moving laboratory" as far as the principle of relativity is concerned? If the plane is *not* jostling (as far as you can tell), is it *certain* that the plane is at rest on the earth's surface? Defend your responses in a short paragraph. (As you write your paragraph, consider how the case of a plane accelerating for takeoff was *rejected* as being a "smoothly moving laboratory" in Sec. 1.4. Can a similar approach be used in this case?)

1.2 GALILEO'S PRINCIPLE. Read Galileo's 1632 presentation of the principle of relativity (the instructor will make a copy available if this problem is assigned). In a short paragraph, compare and contrast his presentation with the presentation given in this chapter. What for Galileo corresponds to a "smoothly moving laboratory"? What for Galileo corresponds to the phrase that the "laws of physics are the same" in such laboratories?

1.3 COMPLETE THE PROOF. Write out a proof for the *second* part of the statement about the relative velocities of inertial frames appearing at the beginning of Sec. 1.5 in the style of the proof of the first part given in the text. That is, show that the definition of an inertial reference frame implies that a rigid reference frame that moves at a constant velocity with respect to any inertial frame must itself be inertial. (Remember that the *definition* of an inertial frame is that a first-law detector registers no violation of Newton's first law. Thus your aim in the proof should be to show that this is true for the reference frame in question. Remember also that the ball in the detector experiences no forces by definition: see Sec. 1.4.)

1.4 ROTATING REFERENCE FRAMES. A reference frame in deep space rotates about its center of mass once in 1.2 h. A first-law detector at its center of mass detects no violation of Newton's first law. A detector located anywhere else in this frame will detect a violation of Newton's first law, because of the apparent "centrifugal force" created by the frame's rotation.

a Estimate the drift acceleration of the ball relative to its detector shell in a first-law detector located 52 m from the center of the frame.

b How long will it take a ball initially at rest in the center of the detector to touch the shell wall (thus registering the violation of the first law) if the shell's inside radius is 1.0 cm greater than the outside radius of the ball?

1.5 FREELY FALLING FRAMES I. A freely falling reference frame near the surface of the earth is not *quite* inertial because the strength of the earth's gravitational field is not uniform across the frame. For example, imagine a rigid reference frame equipped with first-law detectors that is freely falling in space 320 km above the earth's surface. Detector A, fixed to the frame's center of mass, registers no violation of Newton's first law, because the detector ball and the frame's center of mass fall with the same acceleration. Now consider a detector B attached to the frame 52 m away from the center in the direction away from the earth. The *shell* of detector B falls with the same acceleration as the frame's center of mass, since it is fixed to the frame's rigid lattice. But the *ball* inside detector B falls with a *smaller* acceleration because it is slightly farther from the earth than the frame's center is (see Fig. 1.10).

a What is the relative acceleration of detector B's shell and ball in this case (i.e., the difference between the acceleration of the ball and the acceleration of the frame as a whole, relative to the earth)? The radius of the earth is about 6370 km, and its mass is 6.0×10^{24} kg. The universal gravitational constant is $G = 6.67 \times 10^{-11}$ N $\cdot$ m^2/kg^2. [*Hint:* You will find the following approximation useful: $(1 + x)^{-2} \approx 1 - 2x$, if $x << 1$. You can test this approximation on your calculator if you like: try $x = 0.3$, then $x = 0.1$, then $x = 0.03$, noticing how the approximation improves as x gets smaller.]

b How long will it take a ball initially at rest in the center of the detector to touch the shell wall (thus registering the violation of Newton's first law) if the shell's inside radius is 1.0 cm greater than the outside radius of the ball?

FIGURE 1.10
The ball in detector B falls a bit more slowly than the reference frame as a whole because it is farther from the earth than is the center of mass of the reference frame. Thus it will appear to accelerate toward the detector shell as shown.

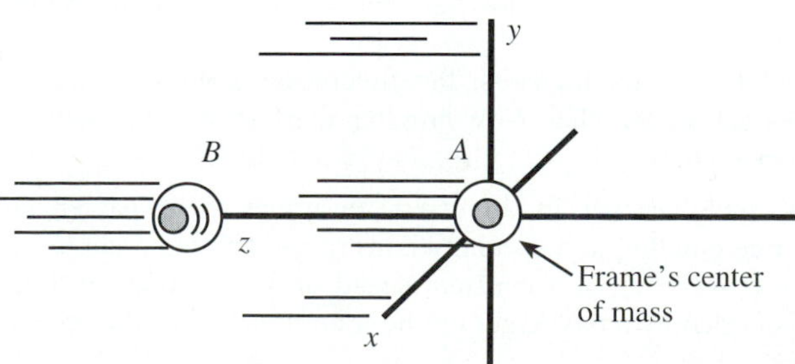

1.6 FREELY FALLING FRAMES II. Do Prob. 1.5, but assume that the frame is 384,000 km from the center of the earth (about the same distance as the moon).

1.7 ORBITING FRAMES. A worried student writes: "I do not see how a freely falling frame can possibly be inertial. Consider a freely falling frame S in orbit around the earth. Consider another frame S' in deep space far away from the earth and essentially at rest with respect to it. According to the definition we are using, both frames are supposed to be inertial. Yet it is clear that orbiting frame S does not move with a constant velocity with respect to the stationary frame S', as required by the argument in Sec. 1.5. Thus either the definition is absurd or the argument must be wrong!" Write a response to this objection describing the error in the student's reasoning. (*Hint:* The argument in Sec. 1.5 assumes that the lattices of the two reference frames in question overlap sufficiently in some region to measure the motion of some common detector ball. Consider extending the lattices of the two frames described in this problem sufficiently far toward each other so that they overlap. Could such extended frames possibly be considered inertial? Why not? Consider the issues discussed in Prob. 1.5.)

1.8 OTHER KINDS OF FIRST-LAW DETECTORS. Design a first-law detector that does *not* use a floating ball as the moving part. Your detector should primarily test Newton's *first* law and not some other law of physics, though you may employ other physical laws in conjunction with a test of Newton's first law. (*Hints:* Any first-law detector must involve the use of some freely moving particle or particles, since it is the motion of such particles that Newton's first law describes. The particles in an ideal gas are essentially freely moving between collisions. Can a box of gas be used as a first-law detector? If so, how? How about using the particles emitted by some radioactive source, or electrons emitted by an electron gun? There are in fact many methods that might be employed. Use your imagination!)

1.9 COORDINATES IN NEWTONIAN RELATIVITY. A train moving with a speed of 55 m/s passes through a railroad station at time $t = t' = 0$. Fifteen seconds later a firecracker explodes on the track 1.0 km away from the train station in the direction that the train is moving. Find the coordinates of this event in both the station frame (consider this to be the Home Frame) and the train frame. Assume that the direction of motion of the train defines the $+x$ direction in both frames.

1.10 COORDINATES IN NEWTONIAN RELATIVITY. Two firecrackers explode simultaneously 125 m apart along the x axis of a certain inertial frame (the Home Frame). Consider a second inertial frame (the Other Frame) in standard orientation with respect to the Home Frame moving at a speed of 25 m/s with respect to it.

a According to the galilean transformation equations, are the firecracker explosion events simultaneous in the Other Frame? How far apart are the explosions as measured in that frame?

b Assume that instead of being simultaneous, the firecracker farthest in the $+x$ direction explodes 3.0 s before the other. Now how far apart are the explosions as measured as in the Other Frame?

1.11 RELATIVE VELOCITY. Imagine that in an effort to attract more passengers, Amtrak trains now offer free bowling in a specially constructed "bowling alley" car. Imagine that such a train is traveling at a constant speed of 35 m/s relative to the ground. A bowling ball is hurled by a passenger on the train in the same direction as the train is traveling. The ball is measured in the ground frame to have a speed of 42 m/s. What is its speed in the frame of the train? (*Hints:* Draw a picture. How should you define the $+x$ direction? Which frame shall we make the Home Frame? Which frame shall we make the Other Frame?)

1.12 RELATIVE VELOCITY. Imagine the situation described in Prob. 1.11 with the following modification. Say that we know that the ball's velocity is 8 m/s in the train frame. What will be its speed relative to the ground?

1.13 RELATIVE VELOCITY. In a certain particle accelerator experiment, two subatomic particles A and B are observed to fly away in opposite directions from a particle decay. Particle A is observed to travel with a speed of $0.6c$ relative to the laboratory, and particle B is observed to travel with a speed of $0.9c$, where c is the speed of light (3.0×10^8 m/s). According to the galilean velocity transformation equations, what speed would particle B be measured to have in an inertial frame attached to particle A? (For simplicity's sake, you can take the direction of motion of particle B to define the $+x$ direction in both the laboratory and particle A frames of reference.) (*Hint:* Take the reference frame of particle A to be the Home Frame and the laboratory frame to be the Other Frame. Draw a picture!)

1.14 NEWTONIAN RELATIVITY. Some people play a game of shuffleboard on an ocean cruiser moving down the Hudson River at a constant speed of 17 m/s in the $+x$ direction relative to the shore. During one shot, a puck (which has a mass m = 750 g and is traveling at 10 m/s in the $-x$ direction in the boat frame) hits a puck having the same mass at rest. After the collision, the first puck comes to rest and the other puck travels at 10 m/s in the $-x$ direction.

a Show that the total momentum of the two-puck system is conserved in the boat frame.

b Imagine that someone sitting on a bridge under which the boat is passing takes a video of this important game. What velocity will each puck be measured to have relative to the shore? Show that in spite of the fact that the puck's x velocities have signs and magnitudes that are *different* from those measured on the boat, the total momentum of the two-puck system is *still* conserved in the shore frame.

1.15 NEWTONIAN RELATIVITY. Imagine two inertial reference frames in standard orientation, the Other Frame moving with a speed β in the $+x$ direction relative to the Home Frame. Consider an arbitrary collision as observed in the Home Frame: an object with mass m_1 and velocity $\vec{v}_1$ hits an object of mass m_2 and velocity $\vec{v}_2$. After the collision, the objects move off with velocities $\vec{v}_3$ and $\vec{v}_4$, respectively. Do *not* assume that all or even any of these velocities are in the x direction. Assume though, that total momentum *is* measured to be conserved in the Home Frame, i.e., that

$$m_1\vec{v}_1 + m_2\vec{v}_2 = m_1\vec{v}_3 + m_2\vec{v}_4 \qquad \text{(assume this!)}$$

Using this equation and the galilean velocity transformation equations, show that the total momentum of the two objects is also conserved in the Other Frame, i.e., that

$$m_1\vec{v}_1' + m_2\vec{v}_2' = m_1\vec{v}_3' + m_2v_4' \qquad \text{(prove this!)}$$

even though the velocities as measured in the two frames are very different.

1.16 NEWTONIAN RELATIVITY. A person in an elevator drops a ball of mass m from rest from a height h above the elevator floor. The elevator is moving at a constant speed β downward with respect to its enclosing building.

- **a** How far will the ball fall in the building frame before it hits the ground? (*Hint:* $>h$!)
- **b** What is the ball's initial vertical velocity in the building frame? (*Hint:* Not 0!)
- **c** Use the law of conservation of energy in the building frame to compute the ball's final speed (as measured in that frame) just before it hits the elevator floor.
- **d** Use the galilean velocity transformation equations and the result of part c to find the ball's final speed in the elevator frame.
- **e** Use the result of part d to show that the law of conservation of energy holds in the elevator frame (assuming that it holds in the building frame).

2

SYNCHRONIZATION, UNITS, AND SPACETIME DIAGRAMS

Absolute, true, and mathematical time, of itself, and from its nature, flows equably without relation to anything external. . . .

Isaac Newton[1]

Newton, forgive me; you found the only way which, in your age, was just about possible for a man of highest thought and creative power. The concepts that you created are even today still guiding our thinking in physics, although we now know that they will have to be replaced by others further removed from the sphere of immediate experience, if we aim at a more profound understanding of relationships.

Albert Einstein[2]

2.1 OVERVIEW

To Newton and his followers, time was self-evidently universal and absolute, flowing "equably without relation to anything external." This makes time a *frame-independent* quantity: every observer in every reference frame (inertial or not!) should be able to agree on what time it is. This assumption about time was considered "self-evident" because it is consistent with our immediate experience about the way that time works.

[1]From Mott's translation (1729) of Newton's *Principia Mathematica* (1686), as revised by Cajori and published by the University of California Press (Berkeley: 1934). The quote is from p. 6 of that edition.
[2]From P. A. Schilpp (ed.), *Albert Einstein: Philosopher-Scientist,* New York: Tudor, 1949, pp. 31–32.

We saw in the last chapter that this assumption directly implies that if an object is measured to have a velocity $\vec{v}$ in one inertial reference frame (the Home Frame), its velocity $\vec{v}'$ measured in another inertial frame (the Other Frame) is given by the vector equation

$$\vec{v}' = \vec{v} - \vec{\beta} \tag{2.1}$$

where $\vec{\beta}$ is the velocity of the Other frame with respect to the Home Frame: this is the vector equivalent of the galilean velocity transformation equations (Eqs. (1.3)]. This equation is a direct consequence of the definition of an inertial reference frame and the assumption that time is frame-independent.

In this chapter we will see that light does not obey this equation! A wide variety of experiments confirm that the velocity of light apparently has the magnitude $c = 2.997925 \times 10^8$ m/s in *all* inertial reference frames, in contradiction to Eq. (2.1). In this chapter, we will review the evidence for this outrageous assertion and begin to wrestle with its consequences. If Eq. (2.1) is wrong, there must be something wrong with our assumptions about time. How can we even define time if it is not universal and absolute?

2.2 THE PROBLEM OF ELECTROMAGNETIC WAVES

In 1873, James Clerk Maxwell published a set of equations (now called *Maxwell's equations*) that summarized the laws of electromagnetism in compact and elegant form. These equations were the culmination of decades of intensive and ingenious research by a number of physicists into the connection between the phenomena of electricity and magnetism, and they represent one of the greatest achievements of nineteenth century physics.

Among the many fascinating consequences of these equations was the prediction that traveling waves could be set up in an electromagnetic field, much like ripples could be produced on the surface of a lake. The speed at which such waves would travel is *completely determined* by various universal constants appearing in the equations (constants whose values were fixed by experiments involving electrical and magnetic phenomena and whose values were fairly well known in 1873): the predicted speed of such waves turns out to be about 3.00×10^8 m/s. Light was already known to be a wave (as the result of research work done in the early nineteenth century by Thomas Young and Augustin Jean Fresnel) that traveled roughly this speed (as measured by Ole Rømer in 1675 and Jean Bernard Léon Foucault in 1846). On the basis of this information, Maxwell asserted that light consisted of such electromagnetic waves. This assertion was supported by later experiments confirming that the value of the speed of light was indeed indistinguishable from the value predicted on the basis of the constants in Maxwell's equations, and particularly by the work of Heinrich Hertz, who was able to directly generate low-frequency electromagnetic waves (i.e., radio waves) and demonstrate that they had the properties predicted by Maxwell's equations.

In short, Maxwell's equations of electromagnetism predicted that light waves must travel at a specific speed $c = 2.997925 \times 10^8$ m/s. The question is, relative to what? The consensus in the physics community at the time (one that Maxwell shared) was that electromagnetic waves were oscillations of a hypothetical medium called the *ether,* just as sound waves are oscillations in air and water waves are oscillations in the surface of a body of water. It was therefore assumed that light waves would travel at the predicted speed c relative to this ether and therefore would be measured to have speed c in a reference frame at rest with respect to the ether.

In all other inertial reference frames, however, light waves must be observed to travel at a speed *different* from c. To see this, imagine a spaceship flying away from a space station at a speed β. A blinker on the space station emits a pulse of light waves toward the departing spacecraft (see Fig. 2.1). Let us imagine that the space station is at rest with respect to the ether and treat this as the Home Frame. Observers on the space station will then measure the emitted pulse of light to move away from the blinker at a speed of c. How rapidly would the pulse of light waves from the blinker be observed to travel in a frame fixed to the spaceship?

The answer, according to the galilean velocity transformation equations, is simple. By construction, both the flash and the spacecraft move in the $+x$ direction. The x component of the galilean velocity transformation equation [Eq. (2.1)] then reads

$$v'_{\text{light},x} = v_{\text{light},x} - \beta = c - \beta \tag{2.2}$$

This means that the speed of the light flash, as measured in the frame of the spacecraft (i.e., the Other Frame), is $c - \beta$. This makes sense. If the spacecraft happened to travel at the speed of light (so that $\beta = c$), the flash should *intuitively* appear to be motionless in the frame of the spacecraft and thus never catch up with it, in agreement with Eq. (2.2). In any frame moving with respect to the ether, then, light waves would be measured to have a speed different from c.

But this means that Maxwell's equations strictly apply only in a certain inertial reference frame (the frame at rest with respect to the ether), since they state that light waves move with a specific speed c. Presumably some small modifications would have to be made to these equations to make them work in frames *not* at rest with respect to the ether.

Now, the ether concept proved to be problematic from the beginning. This ether had to fill all space and permeate all objects and yet be virtually undetectable. It had to

FIGURE 2.1
A light flash chasing a departing spacecraft.

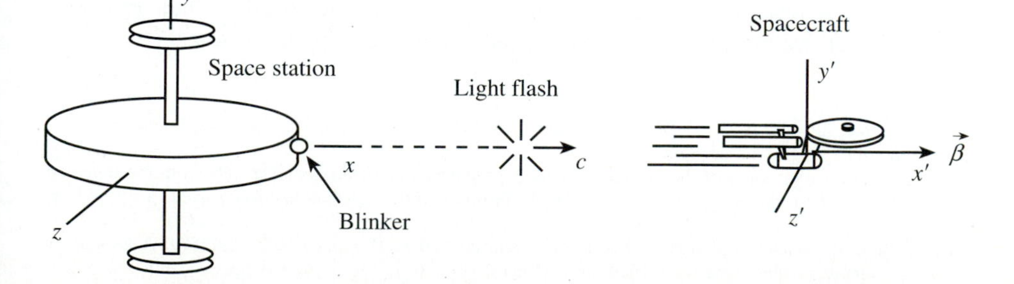

have virtually zero density and viscosity, because it was not observed to significantly impede any object's motion. On the other hand, it had to be extraordinarily "stiff" with regard to oscillations because generally the speed of waves in a medium increases with the stiffness of the medium, and c is very large (for comparison, mechanical waves traveling through solid rock have speeds of only 6000 m/s).

In 1887, American physicists Albert Michelson and Edward Morley performed a sensitive experiment designed to prove the existence of this seemingly indiscernible ether. If this ether filled all space, the earth must (as a result of its orbital motion around the sun) be moving through the ether at a speed comparable to its orbital speed of about 30 km/s. This "ether wind" would have the effect of making the speed of light depend on the direction of its travel: a light wave moving against the ether wind would move more slowly than a wave moving across the wind. Michelson and Morley therefore constructed a very sensitive experiment that compared the speed of two beams of light sent in perpendicular directions. The presence of an ether wind would manifest itself in a slight difference between the speeds of light in these two directions.

To the surprise of everyone involved, it turned out that there was *no* discernible difference in the speeds of the two light waves. The experiment was repeated with different orientations of the apparatus and at different times of the year (just in case the earth happened to be at rest with respect to the ether at the time of the first experiment), and by other physicists. In all cases, the result was that the speed of light was apparently a fixed value, independent of the motion of the earth.

(In an experiment in the 1930s, R. J. Kennedy and E. M. Thorndike showed that although the relative speed of the earth's reference frame in July and its frame in January is roughly 60 km/s, the *numerical value* of the speed of light differed by less than 2 m/s during this time span. This affirms the constant value of the speed of light with a fractional uncertainty of less than 10^{-8}, that is, to better than eight decimal places.[1])

The Michelson-Morley result (and other corroborating results) caused a ruckus in the physics community, as physicists strove to explain away these results while saving the basic ether concept. Many explanations were offered (some involving bizarre effects of the ether wind on measurement devices), but none provided a satisfactory explanation of all known experimental data.[2]

In 1905, Albert Einstein published a short paper on the subject (in the European journal *Annalen der Physik*)[3] that changed the direction of physics. In that paper, Einstein proposed that since it seemed to be impossible to demonstrate the existence of the ether, we should reject the concept of the ether and we simply *accept* the idea that light is able to move in a vacuum. But since the vacuum of empty space provides no anchor defining a special "rest" frame where the speed of light is c (what would such a frame be "at rest" with respect to?), if we accept that the speed of light in a vacuum is c, then we must accept that this speed is c in *every* inertial reference frame, in

[1] *Phys. Rev.,* **42:**400, 1932. See E. F. Taylor and J. A. Wheeler, *Spacetime Physics,* San Francisco: Freeman, 1966, for a lengthy description.

[2] An excellent discussion of the variety of models proposed and experimental results from the time can be found in the first chapter of R. Resnick's book, *Introduction to Special Relativity,* New York: Wiley, 1968.

[3] An English translation of part of this paper is reprinted in A. P. French (ed.), *Einstein: A Centenary Volume,* Cambridge, Mass.: Harvard, 1979, pp. 281–293.

direct contradiction to Eq. (2.2). The assumption *that the speed of light is a frame-independent quantity* is necessary, Einstein argued, to make Maxwell's equations consistent with the principle of relativity, since Maxwell's equations are laws of physics that predict a specific and well-defined value for the speed of light.

The frame independence of the speed of light goes deeply against our intuition. How can a pulse of light waves be measured to move at a speed c in the frame of the space station of our previous example and *also* in the frame of the spacecraft, when the two frames are not at rest with respect to each other? This bold assumption, while neatly sidestepping the experimental difficulties associated with the ether concept, also seemed impossible to most of Einstein's contemporaries.

But there are really only three possibilities: either the principle of relativity is wrong, Maxwell's equations are wrong, or the galilean velocity transformation equations are wrong. By accepting the ether concept, physicists before Einstein had opted to accept the idea that Maxwell's equations would have to be modified in frames moving with respect to the ether, thus keeping the galilean velocity transformation and implicitly rejecting the full principle of relativity. Even as evidence against the ether hypothesis became firm and incontrovertible, rejection of the galilean velocity transformation, so solidly based on simple and obvious ideas, seemed absurd.

On the other hand, what Einstein suggested did have a certain simplicity and elegance. Throw away the ether idea, he said. It is an unhelpful, ad hoc hypothesis with no experimental backing. Throw away the awkward and bizarre theories that arose to explain our inability to detect the ether. Embrace instead the beautiful simplicity of the principle of relativity *and* Maxwell's equations. The speed of light then is the same in all inertial reference frames as a matter of course, and the null results of experiments like the Michelson-Morley experiment are trivially explained.

The cost? *The galilean transformation equations must be wrong.* But what could *possibly* be wrong with their derivation? It is the idea of universal and absolute time that is wrong, replies Einstein. Let us look at what the principle of relativity requires of time.

2.3 RELATIVISTIC CLOCK SYNCHRONIZATION

Our basic problem is that the concept of universal and absolute time seems to be at odds with having the principle of relativity apply to Maxwell's equations, yet experimental results support the idea that the principle *does* apply to Maxwell's equations. If time is not universal and absolute, how can we even define what time means?

The solution, as Einstein was the first to see, is that we must define what we mean by time *operationally* within each inertial frame by specifying a concrete procedure for synchronizing that frame's clocks in a manner consistent with the principle of relativity.

But how can we synchronize clocks in a manner consistent with the principle of relativity? Here is a simple method. Maxwell's equations imply that light moves through a vacuum at a certain fixed speed c. The principle of relativity requires that this speed be the same in every inertial reference frame, as we discussed in Chap. 1. Therefore, *any* method of synchronization that is consistent with the principle of relativity will lead to light being measured to have the speed c in any inertial reference frame.

Since the speed of light has to be c in every inertial frame anyway, let us in fact synchronize the clocks in our inertial reference frame by *assuming* that light always has the same speed of c! How do we do this? Imagine that we have a master clock at the spatial origin of our reference frame. At exactly $t = 0$, we send a light flash from that clock to the other clocks in the frame. Since light is assumed to travel at a speed of $c = 299{,}792{,}458$ m/s, this flash will reach a lattice clock exactly 1.0 m from the master clock at a time of exactly $t = (1.0 \text{ m})/(299{,}792{,}458 \text{ m/s}) = 3.33564095 \times 10^{-9}$ s = 3.33564095 ns. Therefore, if we set this clock to read 3.33564095 ns exactly as the light flash passes, we know that it is synchronized with the master clock. The process is similar for all of the other clocks in the lattice.

Definition of Synchronization (Draft)

A light flash is emitted by clock A in an inertial frame at time t_A (as read on clock A) and received by clock B in the same frame at time t_B (as read on clock B). These clocks are defined to be **synchronized** if $c(t_B - t_A)$ is equal to the distance between the clocks. That is, the clocks are synchronized if they measure the speed of a light signal traveling between them to be c.

For example, imagine that we have a clock on the earth and a clock on the moon. How can we tell if these clocks are synchronized according to this definition? The distance between the earth and the moon is 384,000,000 m, so it should take light 1.28 s to travel from the earth to the moon. If we send a light signal from the earth clock at exactly noon and if the moon clock reads 1.28 s after noon when the signal passes it, the earth and moon clocks are synchronized.

"Now, wait a minute!" I hear you cry. "Isn't this a bit circular? You claim that Maxwell's equations predict that light will be measured to have the same speed c in every inertial reference frame. But then you go setting up the clocks so that this result is *assured.* Is this fair?"

This *is* fair. We are *not* trying here to prove that Maxwell's equations obey the principle of relativity, we are *assuming* they do so that we can determine the *consequences* of this assumption. To make this clear in his original paper, Einstein actually stated the frame independence of the speed of light as a separate postulate, *emphasizing* that it is an assumption. The point is that if the principle of relativity is true, the speed of light will be measured to have the speed c no matter *what* valid synchronization method we use, so why not use a method based on that fact?

Moreover, there *are* other valid methods of synchronizing the clocks in a given inertial reference frame, methods that make no assumptions whatsoever about the frame independence of the speed of light.[1] If such a method were used to synchronize

[1]One of the simplest is described by Alan Macdonald, *Am. J. Phys.*, **51**(9), 1983. Macdonald's method is as follows. Assume that clocks A and B emit flashes of light toward each other at $t = 0$ (as read on each clock's *own* face). If the readings of the two clocks also agree when they *receive* the light signal from the other clock, they are synchronized. This method only assumes that the light flashes take the same time to travel between the clocks in each direction (i.e., there is no preferred *direction* for light travel); it does not assume light has any frame-independent speed. Achin Sen [*Am. J. Phys.*, **62**(2), 1994] presents a particularly nice example of an approach that sidesteps the synchronization issue, showing mathematically that the results of relativity follow directly from the principle of relativity. Sen's article also contains an excellent list of references.

the clocks in an inertial frame, such a frame could be used to verify *independently* that the speed of light is indeed frame-independent. Such alternative methods have been shown to yield the same consequences as one gets assuming the frame independence of the speed of light. But these methods are also more complex and abstract: the definition of synchronization in terms of light is much more vivid and easy to use in practice.

2.4 MEASURING DISTANCE IN TIME UNITS

In ordinary SI units, the speed of light c is equal to 299,792,458 m/s, a somewhat ungainly quantity. The definition of clock synchronization given in Sec. 2.3 means that we will often need to calculate how long it would take light to cover a certain distance or how far light will travel in a certain time. You can perhaps see how messy such calculations will be.

For this reason and many others it will be convenient when studying relativity theory to measure distance not in the conventional unit of meters but in a new unit called a *light-second,* or just *second* for short. A **light-second** is defined to be the distance that light travels in 1 second of time. Since 1983, the meter has in fact been officially *defined* by international agreement as the distance that light travels in 1/299,792,458 second. Therefore there are exactly 299,792,458 meters in 1 light-second *by definition.*

We can, of course, measure distance in any units that we please; there is nothing magically significant about the meter. But choosing to measure distance in light-seconds has some important advantages. In the first place, light travels exactly 1 second of distance in 1 second of time *by definition.* This allows us to talk about clock synchronization much more easily. For example, if clocks A and B are 7.3 light-seconds apart in an inertial reference frame and a light flash leaves clock A when it reads $t_A = 4.3$ s, the flash should arrive at clock B at a time $t_B = (4.3 + 7.3)$ s $=$ 11.6 s if the two clocks are correctly synchronized, since light travels exactly 1 light-second in 1 second of time by definition. You can see that there are no ungainly unit calculations to perform if we measure distance this way.

Indeed, agreeing to measure distance in seconds allows us to state the definition of clock synchronization in an inertial frame in a particularly nice and concise manner:

Definition of Synchronization (Final Version)

Two clocks in an inertial reference frame are defined to be **synchronized** if the time interval (in seconds) registered by the clocks for a light flash to travel between them is equal to their separation (in light-seconds).

In spite of the tangible advantages that measuring distance in seconds yields when it comes to talking about synchronization, this is not the most important reason for choosing to do so. In the course of working with relativity theory, we will uncover a deep relationship between time intervals and distance intervals, akin to the relationship between distances measured north and distances measured east on a plane. One would not think of measuring northward distances in feet, say, and eastward distances in meters: that would obscure the fundamental similarity of these measurements. Similarly, measuring time intervals in seconds while measuring distance intervals in meters obscures the fundamental similarity in these measurements that will be illuminated by relativity theory. Choosing to measure time in the same units as distance will make this beautiful symmetry of nature more apparent.

2.5 THE SR SYSTEM OF UNITS

The standard unit system used by scientists studying ordinary phenomena is the *Système International,* or SI, unit system. In this system, the units for mechanical quantities (velocity, momentum, force, energy, angular momentum, pressure, and so on) are based on three fundamental units: the *meter,* the *second,* and the *kilogram.* In this text, however, we will use a slightly modified version of the SI unit system (let us call it the *Système Relativistique,* or SR, unit system), where distance is measured in *seconds* (i.e., light-seconds) instead of in *meters* (with the other basic units being the same).

The standard method of converting *any* quantity from one kind of unit to another is the method of *conversion factors.* One first writes down an equation stating the basic relationship between the units in question: 1 mi = 1.609 km, for example. Such an equation can be stated in the form of a ratio equal to 1: 1 = 1 mi / 1.609 km or 1 = 1.609 km / 1 mi. Since multiplying by 1 does not change a quantity, you can multiply the original quantity by whichever ratio leads you to the correct final units upon cancellation of any units that appear in both the numerator and denominator. For example, to determine how many miles there are in 25 km, you simply multiply the 25 km by the first of the two ratios described above, as follows:

$$25 \text{ km} = 25 \text{ km} \cdot 1 = 25 \cancel{\text{km}}\left(\frac{1 \text{ mi}}{1.609 \cancel{\text{km}}}\right) = \frac{25}{1.609} \text{ mi} \approx 16 \text{ mi} \tag{2.3}$$

The factors used to convert from SI distance units to SR distance units are based on the fundamental definition of the light-second: 299,792,458 m $\equiv$ 1 s. Thus the basic conversion factors are 1 = 1 s / 299,792,458 m or 1 = 299,792,458 m / 1 s. For example, a distance of 25 km can be converted into light-seconds as follows:

$$\begin{aligned} 25 \text{ km} &= 25 \text{ km} \cdot 1 \cdot 1 = 25 \cancel{\text{km}}\left(\frac{1000 \cancel{\text{m}}}{1 \cancel{\text{km}}}\right)\left(\frac{1 \text{ s}}{2.998 \times 10^8 \cancel{\text{m}}}\right) \\ &\approx 8.3 \times 10^{-5} \text{ s} \approx 83 \ \mu\text{s} \end{aligned} \tag{2.4}$$

meaning that 25 km is equivalent to the distance that light travels in 83 millionths of a second.

The light-second is a rather large unit of distance (the moon is only about 1.3 light-seconds away from the earth!). The light-nanosecond $\equiv 10^{-9}$ light-second (= 0.2998 m $\approx$ 1 ft) is a more appropriate unit on the human scale. On the astronomical scale, the light-year $\approx 3.16 \times 10^7$ s $\approx 0.95 \times 10^{15}$ m is the appropriate (and commonly used) distance unit. The dimensions of the solar system are conveniently measured in light-hours (the solar system is about 10 light-hours in diameter). All these units represent extensions of the basic unit of the light-second.

In the SR unit system, the light-second is considered to be *equivalent* to the second of time, and both units will be simply referred to as *seconds*. This means that these units can be canceled if one appears in the numerator of an expression and the other appears in the denominator. For example, velocity in the SI system is expressed in units of meters per second, but in the SR system, velocities are expressed in seconds per second = unitless(!). Thus an object that travels 0.5 light-second in 1.0 s has a speed in the SR system of 0.5 s / 1.0 s = 0.5 (no units!). The bare number representing a speed in this case actually represents a comparison of the particle's speed to the speed of light, since light covers 1.0 second of distance in 1.0 second of time by definition. Thus a particle traveling at a speed of 0.5 (in SR units) is traveling at half the speed of light.

In a similar manner, one can find the natural units for any physical quantity in the SR system. For example, the kinetic energy of a particle has units of mass·$(\text{speed})^2$. In the SI system, these units would be kg·m^2/s^2. Thus the natural unit of energy in this system is the joule, which is defined to be 1 kg·m^2/s^2. In the SR system, mass·$(\text{speed})^2$ has units of kg·s^2/s^2 = kg. Thus the natural SR unit of energy is the kilogram (the same as the unit of mass!). But how much energy is represented by an SR kilogram of energy? We can determine this by using the standard conversion factor to convert the SR distance unit of seconds to the SI distance unit of meters:

$$1 \text{ kg of energy} = 1\,\cancel{\text{kg}}\frac{\cancel{\text{s}^2}}{\cancel{\text{s}^2}}\left(\frac{2.998 \times 10^8\,\cancel{\text{m}}}{1\cancel{\text{s}}}\right)^2\left(\frac{1\text{ J}}{1\,\cancel{\text{kg}}\cdot\cancel{\text{m}^2}/\cancel{\text{s}^2}}\right) = 8.988 \times 10^{16}\text{ J} \quad (2.5)$$

This unit is roughly equal to the yearly energy output of 10 full-sized electrical power plants!

The general trick when converting from SR units to SI units is to multiply the SR quantity by the appropriate power of the conversion factor 1 = 299,792,458 m / 1 s that yields the correct power of meters in the units of the final result. Similarly, to convert from SI units to SR units, multiply the SI quantity by whatever power of this factor causes the units of meters to disappear (see Appendix A for a complete discussion of how to convert units and equations from one unit system to the other).

2.6 SPACETIME DIAGRAMS

The clock synchronization method described in Sec. 2.3 completes the description of how to build and operate an inertial reference frame. We now know exactly how to assign spacetime coordinates to any event occurring within that inertial frame.

Problems in relativity theory often involve studying the coordinates of events and how events relate to one another. The coordinates of events and the relationships between them can be conveniently and visually depicted using a special kind of graph called a **spacetime diagram.** We will find spacetime diagrams to be nearly indispensable for clearly expressing such relationships.

Consider an event A whose spacetime coordinates are measured in some inertial frame to be t_A, x_A, y_A, z_A. To simplify our discussion somewhat, assume $y_A = z_A = 0$ (that is, the event occurs somewhere along the x axis of the frame). Now imagine

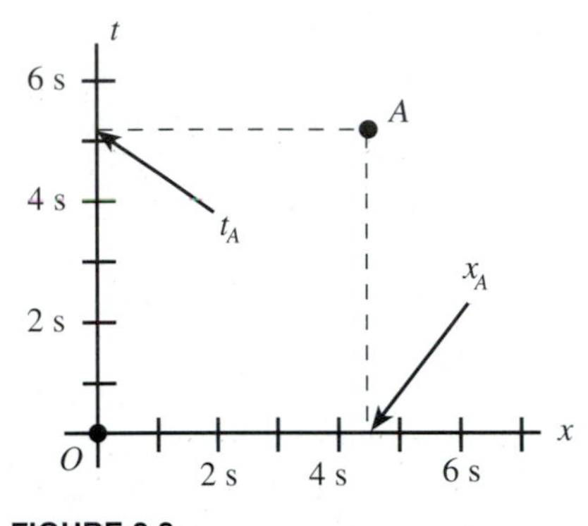

FIGURE 2.2
Plotting event *A* on a spacetime diagram. In the particular case shown, event *A* has a time coordinate $t_A \approx 5.2$ s and an *x* coordinate of $x_A \approx 4.5$ s.

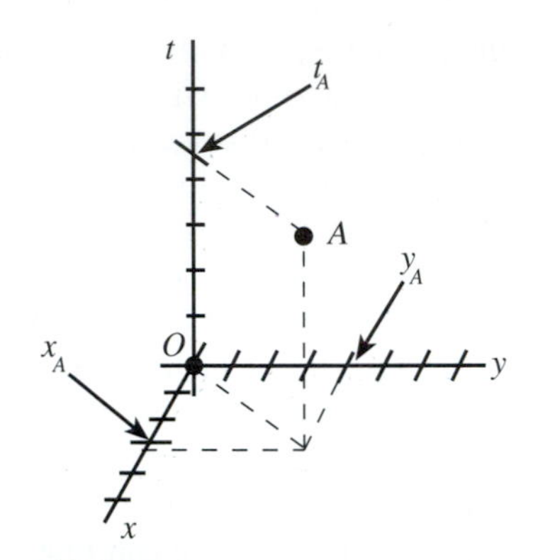

FIGURE 2.3
Plotting an event *A* that has nonzero *x* and *y* coordinates (but a zero *z* coordinate).

drawing a pair of coordinate axes on a sheet of paper. Label the vertical axis with a t and the horizontal axis with an x: it is conventional to take the t axis to be vertical in spacetime diagrams. Choose an appropriate scale for each of these axes. Now you can represent *when* and *where* event A occurs by plotting the event as a point on the diagram (in the usual manner that you would use in plotting a point on a graph) as shown in Fig. 2.2. *Any* event that occurs along the x axis in space can be plotted as a point on such a diagram in a similar manner.

Note that the point marked O on Fig. 2.2 also represents an event. This event occurs at time $t = 0$ and at position $x = 0$. This event is called the **origin event** of the diagram.

If we need to draw a spacetime diagram of an event A that occurs in space somewhere in the xy plane (i.e., which has $z_A = 0$ but nonzero t_A, x_A, y_A), we must add another axis to the spacetime diagram (as shown in Fig. 2.3). The diagram in this case is somewhat less satisfactory and more difficult to draw because we are trying to represent a three-dimensional graph on a two-dimensional sheet of paper.

A spacetime diagram showing events with *three* nonzero spatial components is impossible to draw, as that would involve trying to represent a four-dimensional graph on a two-dimensional sheet of paper. A four-dimensional graph is hard to visualize at all, much less draw! Fortunately, diagrams with one or two spatial dimensions will be sufficient for most purposes.

2.7 THE WORLDLINE OF A PARTICLE

In Sec. 1.2 we visualized describing the motion of an object by imagining the object to carry a strobe light that blinks at regular intervals. If we can specify when and where each of these blink events occurs (by specifying its spacetime coordinates), we can get a pretty good idea of how the object is moving. If the time interval between these blink events is reduced, we get an even clearer picture of the object's motion. In the limit that the interval between blink events goes to zero, the object's motion can be described in unlimited detail by a list of such events. Thus the motion of any object can be described in terms of a connected *sequence* of events. The set of *all* events occurring along the path of a particle is called the **worldline** of the particle.

On a spacetime diagram, an event is represented by a point. Therefore a worldline is represented on a spacetime diagram by an infinite set of infinitesimally separated points, i.e., a curve. This curve is nothing more than a graph of the position of the particle vs. time (except that the time axis is conventionally taken to be vertical on a spacetime diagram). Figure 2.4 illustrates worldlines for several examples of objects moving in the x direction.

Note that because the time axis is taken to be *vertical,* the slope of the curve on a spacetime diagram representing the worldline of an object traveling at constant velocity in the x direction is not v_x (as one might expect) but rise/run $= \Delta t / \Delta x = 1/v_x$! Thus the slope of the curve representing the worldline of an object at rest is infinity and *decreases* as v_x increases. The worldline of an object traveling with a constant x velocity will thus have a constant slope.

Occasionally we will need to draw a spacetime diagram of the worldline of an object moving in two spatial dimensions. The necessary spacetime diagram is then three-dimensional. An example of such a spacetime diagram is shown in Fig. 2.5.

In drawing spacetime diagrams it is also convenient and conventional to use the same-sized scale on both axes. If this is done, the worldline of a flash (i.e., a very short pulse) of light *always* has a slope of either 1 (if the flash is moving in the $+x$ direction) or -1 (if the flash is moving in the $-x$ direction), since light travels 1.0 second of distance in 1.0 second by definition in *every* inertial reference frame. It is also conventional to draw the worldline of a flash of light with a dashed line instead of a solid line (see Fig. 2.6).

FIGURE 2.4
(a) A spacetime diagram displaying the motion of an object by plotting the "blink events" that occur along the path of the object. (b) As the interval between the blink events goes to zero, the set of points representing those events becomes a curve, called a *worldline.* (c) The x velocity of an object moving along the spatial x axis is equal to the *inverse* of the slope of its worldline: slope = rise / run $= \Delta t/\Delta x = 1/v_x$. (d) The left worldline describes an object that starts at rest at $x = 0$ but then travels with ever-increasing speed in the $+x$ direction. The right worldline describes an object at rest.

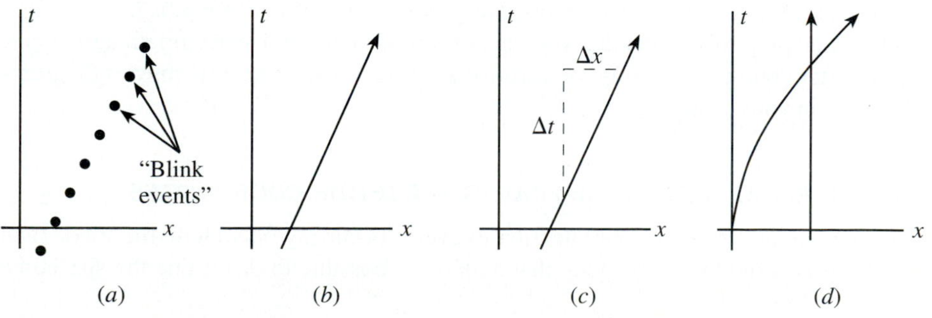

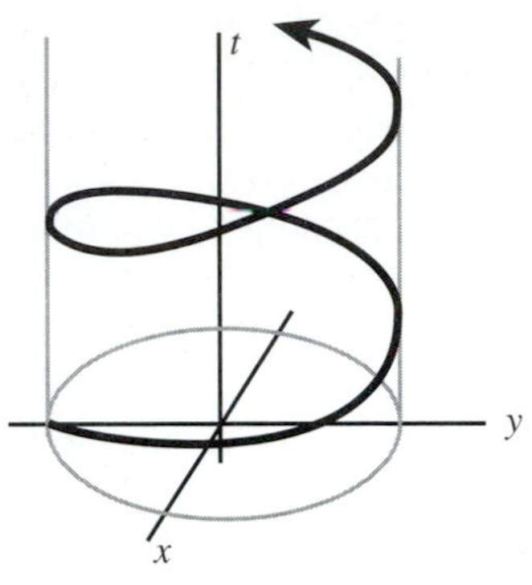

FIGURE 2.5
The worldline of a particle traveling in a circular path about the origin in the *xy* plane.

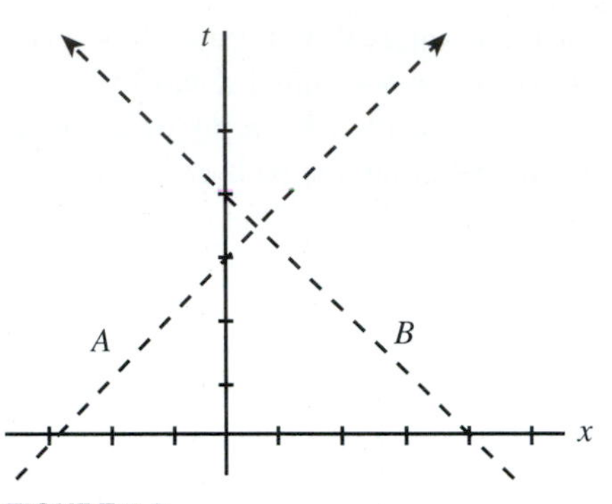

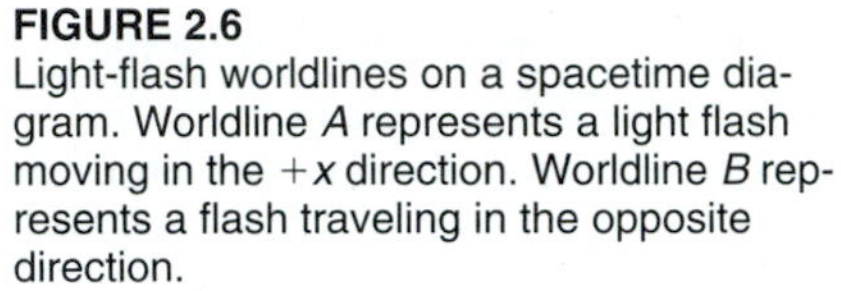
FIGURE 2.6
Light-flash worldlines on a spacetime diagram. Worldline *A* represents a light flash moving in the $+x$ direction. Worldline *B* represents a flash traveling in the opposite direction.

2.8 MAKING A SPACETIME DIAGRAM INTO A "MOVIE"

It is easy to get confused about what a spacetime diagram really represents. For example, in the spacetime diagram shown in Fig. 2.6, it is easy to forget that the light flashes shown are moving in only one dimension (along the x axis), not in two. The velocity vectors of the light flashes shown above point *opposite* to each other, not perpendicular to each other.

There is a technique that you can use to make the meaning of any spacetime diagram clear and vivid: Turn it into a movie! Here is how. Take a piece of card stock and cut a narrow horizontal slit in it with a knife or a razor blade (I suggest making a slit about 1/16 in wide and about 4.5 in long on a 4×6 index card). This slit represents the spatial x axis at a given instant of time. Now place the slit over the x axis of the spacetime diagram. This shows you what is happening along the spatial x axis at time $t = 0$. Now slowly move the slit up the diagram, keeping it parallel to the x axis drawn on the diagram: this will show you what is happening along the spatial x axis at successively later times. You can watch the objects whose worldlines are shown on the diagram "move" to the left or right as the slit exposes different parts of their worldlines. Events drawn as dots on the diagram will show up as "flashes" as you move the slit past them. What you are seeing through the slit as you move it up the diagram is essentially a movie of what happens along the x axis as time passes. The process is illustrated in Fig. 2.7.

If you employ this technique, you cannot fail to interpret a spacetime diagram correctly. After you practice using the card for a bit, you will be able to convert diagrams to movies in your head.

2.9 THE RADAR METHOD OF FINDING SPACETIME COORDINATES

If we are willing to confine our attention to events occurring only along the x axis (and thus to objects moving only along that axis), it is possible to determine the spacetime

FIGURE 2.7
(Read from bottom up)

(*d*)

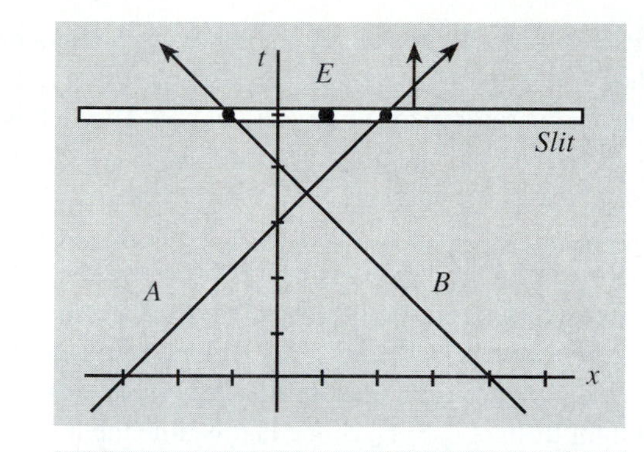

(*d*) Now the light flashes have passed each other and are moving away from each other. You can also see through the slit the momentary flash representing the firecracker explosion (event *E*).

(*c*)

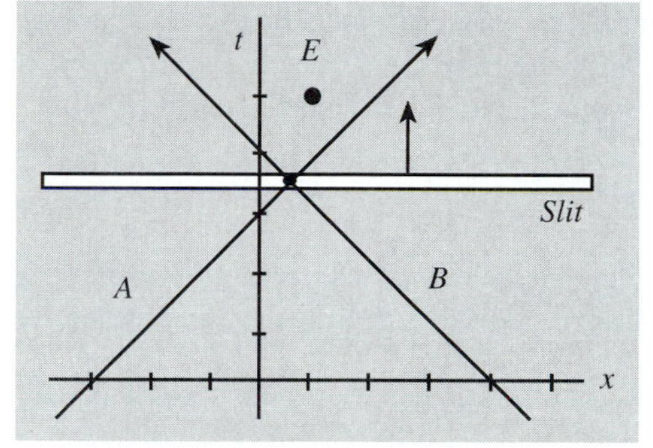

(*c*) At this instant, the light flashes pass through each other at a position of about $x = +1$ m.

(*b*)

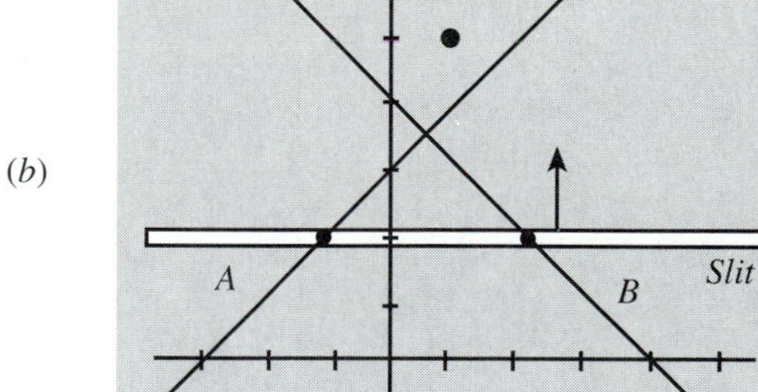

(*b*) As time passes (and you move the slit up the diagram) the dots representing the light flashes approach each other.

(*a*)

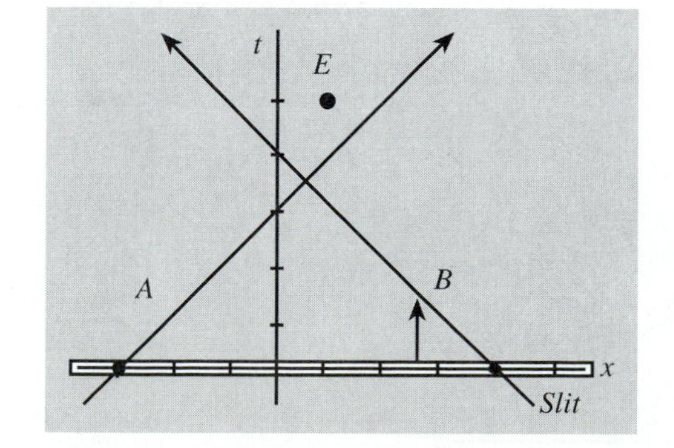

(*a*) The spacetime diagram is basically the same as Fig. 2.6 with a firecracker explosion (event *E*) thrown in to make things more interesting. At time $t = 0$, the light flashes are represented by black dots that are visible through the slit at $x = -3$ m and $x = +4$ m.

coordinates of an event with a *single* master clock and some light flashes: we do not need to construct a lattice at all! The method is analogous to locating an airplane using radar.

Imagine that at the spatial origin of our reference frame (i.e., at $x = 0$) we have a master clock that periodically sends flashes of light in the $\pm x$ directions. Imagine that a certain flash emitted by the master clock at t_A happens to illuminate an event E of interest that occurs somewhere down the x axis. The reflected light from the event travels back along the x axis to the master clock, which registers the reception of the reflected flash at time t_B. This situation is illustrated by the spacetime diagram in Fig. 2.8. The values of the emission and reception times t_A and t_B are sufficient to determine both the location and the time that event E occurred! Consider first how to determine the location. The round-trip time of the light flash is $t_B - t_A$. Since in this time the light covered the distance from $x = 0$ to event E and back and since the light flash travels 1 second of distance in 1 second of time by definition, the distance to event E (in seconds) must be half the round-trip time (in seconds). The x coordinate of event E is thus

$$x_E = \pm \frac{1}{2}(t_B - t_A) \tag{2.6a}$$

the sign being determined by whether the reflected flash comes from the $-x$ or $+x$ direction (in this case, the reflected flash comes from the $+x$ direction, so we would select the positive sign).

We can determine the time coordinate t_E of the event in the following manner. Since the light flash traveled the same distance *to* the event as back *from* the event and since the speed of light is a constant, the event must have occurred exactly halfway between times t_A and t_B. The midpoint in time between times t_A and t_B can be found by computing the average, so

$$t_E = \frac{1}{2}(t_B + t_A) \tag{2.6b}$$

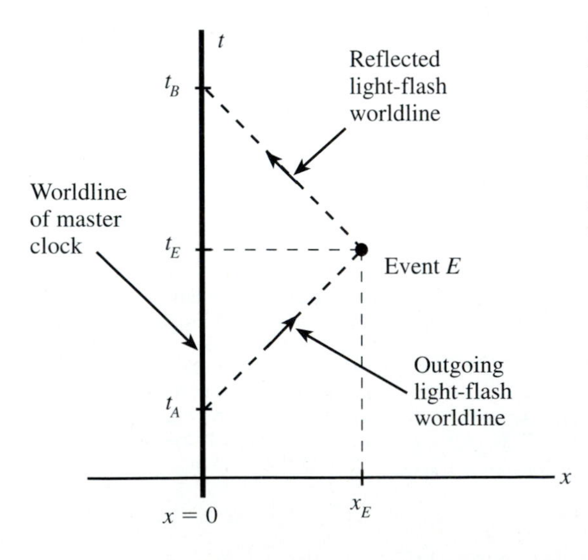

FIGURE 2.8
At time t_A, the master clock at rest at $x = 0$ in the Home Frame sends out a flash of light. This flash reflects from an event E of interest somewhere down the x axis and returns to the master clock at time t_B. What are the spacetime coordinates t_E and x_E of the event E?

Equations (2.6) represent a method of determining the spacetime coordinates of an event that does *not* require the use of a complete lattice of synchronized clocks, but you should be able to convince yourself that *the method illustrated here will produce exactly the same coordinate values that you would get from a lattice.* For example, imagine that you actually had a lattice clock at x_E (the location of event E). The distance between that clock and the master clock at the origin *must* be equal to half the time it would take a flash of light to go from one clock to the other and back, since light travels a second of distance in a second of time by definition. Moreover, at event E the lattice clock at x_E *must* read t_A + (the light travel time between the two clocks) $= t_A + 1/2\,(t_B - t_A) = 1/2\,(t_B + t_A)$, since two clocks are *defined* to be synchronized if they measure a light flash to travel between them at the speed of light (1 second of distance in 1 second of time).

Using the radar method to determine spacetime coordinates is therefore *equivalent* to using a lattice of synchronized clocks. In some cases in this text, we will find it clearer or more convenient to use one method to explain a concept, and in some cases the other.

The radar method is actually used to assign coordinates to events and track the trajectories of aircraft (the impracticality of using a clock lattice to do the same is obvious!) This method can be used to precisely locate an object in *three* spatial dimensions if one not only records the time that the outgoing pulse is sent and the time that the reflected pulse was received but also the *direction* from which the reflected pulse was received (which is usually expressed in the form of two angles: an angle from north in the horizontal plane and an angle up from that plane). The analysis is more complicated, but the basic method is the same.

The important thing to realize is that either the radar method or the clock-lattice method provides specific, well-defined, and equivalent ways of assigning time coordinates to events, and both methods are based on the assumption that the speed of light is a frame-independent constant. These equivalent methods essentially define what time *means* in special relativity and thus will provide the foundation for most of the arguments in the remainder of the text. *Make sure that you thoroughly understand both these methods.*

2.10 SUMMARY AND REVIEW

Let us review the main ideas in this chapter.

1. The theory of electromagnetism (as expressed by Maxwell's equations) predicts the existence of electromagnetic waves that travel with a certain well-defined speed $c = 3.00 \times 10^8$ m/s. These waves are identified as being light.
2. A variety of experimental evidence, gathered mostly since 1850, supports the assertion that Maxwell's equations hold true in all inertial reference frames or, equivalently, that the speed of light has the same value in all inertial reference frames. This is consistent with the principle of relativity but contradicts the galilean velocity transformation.

3 The galilean velocity transformation rests on the assumption that time is absolute and frame-independent. It is specifically this assumption that Einstein called into question.

4 An alternative approach to defining time that is consistent with the principle of relativity is to *define* two clocks in a lattice to be synchronized if they register the speed of a light flash traveling between them to have the speed c. Any definition of time that leads to *different* results would mean that Maxwell's equations are not consistent with the principle of relativity.

5 It is convenient to measure distance in seconds, where 1 second of distance is the distance that light travels in 1 second of time ($\approx$ 300,000 km). The SR unit system is the same as the standard SI unit system except that distance is measured in seconds. In the SR system, velocity is a unitless number (expressing the magnitude of the velocity as a fraction of the speed of light), and mass, momentum, and energy all have the same units of kilograms. The general method of conversion from one unit system to the other involves multiplying or dividing by powers of $1 = 3.00 \times 10^8$ m / 1 s until the units come out right.

6 A spacetime diagram is a useful tool for displaying the spacetime coordinates of events and the motion of objects. In a typical spacetime diagram, one plots an event as a point on a graph, where time is plotted vertically and position along one spatial dimension (usually the x axis) is plotted horizontally. The motion of an object is indicated on such a diagram by a curved line, called a *worldline.* If the scale on both axes is the same (this is the conventional choice), the *inverse* slope of an object's worldline at a given time is equal to the object's x velocity at that time. Light flashes have speed equal to 1, so are plotted as lines having slope ± 1 (that is, lines that make an angle of 45° with respect to the vertical).

7 The radar method of determining spacetime coordinates is equivalent to using a synchronized lattice of clocks (and is often more convenient in practice!). If a light flash sent from $x = 0$ at time t_A reflects from event E somewhere along the x axis and returns to $x = 0$ at time t_B, the spacetime coordinates of the event are

$$x_E = \pm \frac{1}{2}(t_B - \mathrm{t_A}) \qquad \text{and} \qquad t_E = \frac{1}{2}(t_B + \mathrm{t_A}) \tag{2.6}$$

This method can in principle be extended to events not lying on the x axis, but it becomes more complicated.

In short, we have developed a method of synchronizing clocks that is consistent with the principle of relativity. We have learned to measure both time and space in seconds to facilitate working with this synchronization method. Most importantly, we have discovered how to use spacetime diagrams to represent the coordinates of events and the motion of particles in a vivid and accessible manner. We now have the tools to begin unraveling the consequences of the principle of relativity and the relativistic picture of time.

What is the nature of time, now that we have unhooked ourself from the notion of a universal and absolute time? We have defined a procedure to measure the time coordinate of an event. But does this definition mean that time is simply a human mental construct, or does there remain something real and universal about time that does not depend on how we build reference frames? In the next three chapters we will explore the meaning of time in the context of relativity theory.

What we have done so far has been to lay the groundwork for a careful exploration of the principle of relativity. We are now ready to unwrap the package that the principle represents and discover the surprises and beauties within.

PROBLEMS

2.1 SYNCHRONIZATION. A firecracker explodes 30 km away from an observer sitting next to a certain clock A. The light from the firecracker explosion reaches the observer at exactly $t = 0$, according to clock A. Imagine that the flash of the firecracker explosion illuminates the face of another clock B which is sitting next to the firecracker. What time would clock B register at the moment of illumination, if it is correctly synchronized with clock A? Express your answer in seconds.

2.2 SYNCHRONIZATION. Imagine that you are in an inertial frame in empty space with a clock, a telescope, and a powerful strobe light. A friend is sitting in the same frame a very large (unknown) distance from your clock. At precisely 12:00:00 noon according to your clock, you set off the strobe lamp. Precisely 30.0 s later you see in your telescope the face of your friend's clock illuminated by your strobe flash. How far away is your friend from you (in seconds)? What should you see on the face of your friend's clock if that clock is synchronized with yours? Describe your reasoning in a few short sentences.

2.3 SR UNITS

a What is the diameter of the earth in seconds?

b A sign on a hiking trail reads "Viewpoint: 5.5 μs." About how long would it take you to walk to that viewpoint? ($1\ \mu s = 10^{-6}$ s.)

c A sign on the highway reads "Speed Limit 6×10^{-8}," meaning speed in SR units. Translate this into miles per hour.

d It was mentioned in Sec. 1.1 that the *Voyager 2* spacecraft achieved speeds in excess of 72,000 km/h. What is this speed in SR units?

e Show that in the SR system, mass, momentum, and kinetic energy all are measured in kilograms. Imagine a large truck with a mass of 25 metric tons (25,000 kg) barreling down a highway at a speed of 95 km/h ($\approx$ 59 mi/h). What is its momentum in kilograms? Its kinetic energy in kilograms?

f In the SI system, power is measured in watts ($1\ W = 1\ J/s = 1\ kg{\cdot}m^2/s^3$). What is the natural unit of power in the SR system? Let us call this unit a "Superwatt." A large electrical power plant produces roughly 1.0×10^9 W. What is this in Superwatts?

g In the SI system, pressure is normally measured in units of pascals ($1\ Pa \equiv 1\ N/m^2$). What is the natural unit of pressure in the SR system? What is this unit in pascals?

h What is the natural unit of acceleration in the SR system? Express the magnitude of this unit in g's ($1g \equiv$ the acceleration of gravity at the earth's surface $= 9.80\ m/s^2$).

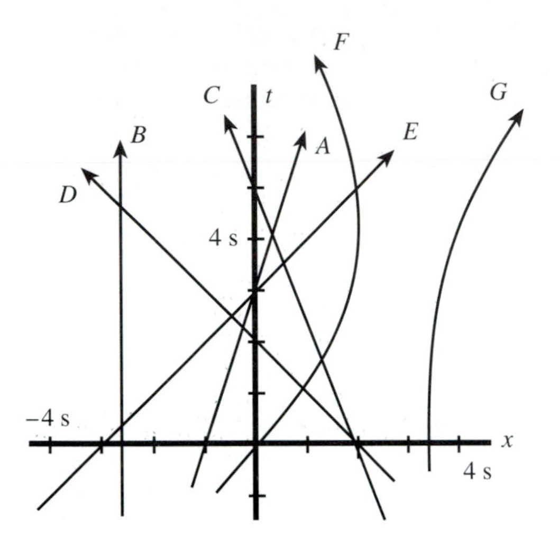

FIGURE 2.9
The worldlines of some sample particles.

2.4 INTERPRETING WORLDLINES. For each of the worldlines shown in the spacetime diagram in Fig. 2.9, describe in words what the particles are doing, giving numerical values for velocities when possible. For example, you might say, "particle A is traveling in the $+x$ direction with a constant speed of 1/3."

2.5 DRAWING WORLDLINES. Draw a spacetime diagram showing the worldlines of the following particles:

a Particle A passes the point $x = 0$ at time $t = 0$ traveling at a constant speed of 3/5 in the $+x$ direction.

b Particle B passes the point $x = 3$ s at $t = 0$ traveling at a constant speed of 1/4 in the $-x$ direction.

c Particle C passes the point $x = 0$ at $t = 2$ s traveling at a speed of 1/2 but constantly decelerates from that speed to rest as time goes on.

d Particle D starts from rest at $x = 0$ at $t = -2$ s and increases speed until it runs into a brick wall at $x = 5$ s and $t = 4$ s, whereupon it stops dead and remains at rest thereafter.

e A flash of light is emitted from position $x = 5$ s at time $t = 1$ s and then travels in the $-x$ direction.

f A flash of light is emitted from the position $x = -2$ s at time $t = 0$ and then moves in the $+x$ direction.

2.6 READING A SPACETIME DIAGRAM. The spacetime diagram in Fig. 2.10 shows the worldline of a rocket as it leaves the earth, travels for a fixed time, and then explodes.

a The rocket leaves earth; the rocket comes to rest in deep space; the rocket explodes. What are the coordinates of each of these three events?

b What is the speed of the rocket relative to the earth during the first phase of its flight?

c A light signal from the earth reaches the rocket just as it explodes. Indicate on the diagram exactly when and where this light signal was emitted.

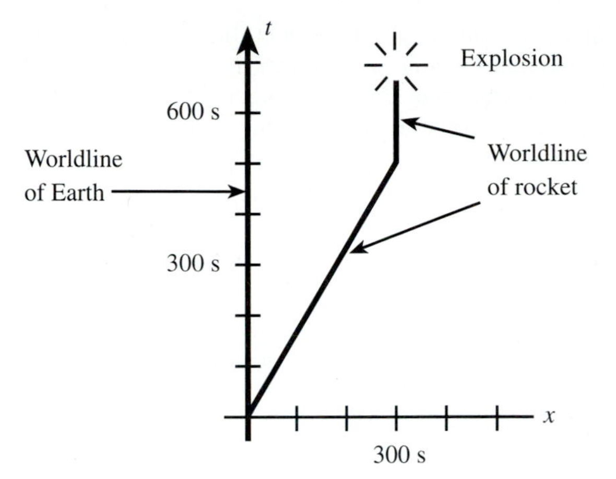

FIGURE 2.10
Spacetime diagram for Prob. 2.6.

2.7 SPACETIME DIAGRAMS. A rocket is launched from the moon and travels away from it at a speed of 2/5. Call the event of the rocket's launching event *A*. After 125 s as measured in the reference frame of the moon, the rocket explodes: call this event *B*. The light from the explosion travels back to the moon: call its reception event *C*. Let the moon itself represent $x = 0$ in its own reference frame, and let event *A* define $t = 0$. Assume that the rocket moves along the $+x$ direction.

a Set up a spacetime diagram of the situation, drawing and labeling the worldlines of the moon, the rocket, and the light flash emitted by the explosion and received on the moon.

b Draw and label the events *A, B,* and *C* as points at the appropriate places on the diagram. Write down the t and x coordinates of these events.

2.8 SPACETIME DIAGRAMS. Imagine that a spaceship in deep space is approaching a space station at a constant speed of $v = 3/4$. Let the space station define the point $x = 0$ in its own reference frame. At time $t = 0$, the spaceship is 16.0 light-hours away from the station. At that time and place (call this event *A*), the spaceship sends a laser pulse of light toward the station, signaling its intention to dock. The station receives the signal at its position of $x = 0$ (call this event *B*), and after a pause of 0.5 h, emits another laser pulse signaling permission to dock (call this event *C*). The spaceship receives this pulse (call this event *D*) and immediately begins to decelerate. It arrives at rest at the space station (call this event *E*) 6.0 h after event *D*, according to clocks in the station.

a Carefully construct a spacetime diagram showing the processes described above. On your diagram, you should show the worldlines of the space station, the approaching spaceship and the two light pulses, as well as indicating the time and place (in the station's frame) of events *A* through *E* by labeling the corresponding points on the spacetime diagram. In particular, exactly when and where does event *D* occur? Event *E?* Write down the coordinates of these events in the station frame. Scale your axes using the hour as the basic time and distance unit, and give your answers in hours.

b Compute the magnitude of the average acceleration of the spaceship between events *D* and *E* in natural SR units (s^{-1}) and in *g*'s ($1g = 9.8$ m/s^2). Note that a shockproof watch can typically tolerate an acceleration of about 50 *g*.

2.9 RADAR METHOD. Imagine that you send out a light flash at time 3.0 s as read by your clock and receive a return reflection showing your brother making a silly face at time 11.0 s as read by your clock. At what time did your brother actually make this silly face? How far is your brother away from you (express in seconds and kilometers)? Is this far enough away that he cannot really be a nuisance?

2.10 RADAR METHOD. Imagine that you send out a radar pulse at time -22 h as measured by your clock and receive a return reflection from an alien spacecraft at time $+12$ h as measured by your clock. Is the spaceship inside or outside the solar system? When did the spaceship reflect the radar pulse?

2.11 RADAR METHOD. Imagine that we want to design an airport radar system that is able to estimate the distances to nearby aircraft to within a few meters. How accurately would such a system have to measure the time between an emitted radar pulse and the reception of the reflection from a given plane?

2.12 "SEEING" IS NOT THE SAME AS "OBSERVING." A bullet train running at high speed passes two trackside signs (*A* and *B*) as shown in the aerial view in Fig. 2.11. Let event *A* be the passing of the *front* end of the train by sign *A* and event *B* be the passing of the *rear* end of the train by sign *B*. An observer is located at the cross marked by an *O* in the diagram. This observer *sees* (i.e., receives light with her eyes) event *A* to occur at time $t = 0$ and *sees* event *B* to occur at time $t = 25$ ns. When does she *observe* these events to occur? That is, what would a clock present at sign *A* read at event *A* and what would a clock present at sign *B* read at event *B* if these clocks were correctly synchronized with the clock at *O*? In what way is the diagram misleading about the implied time relationship between events *A* and *B*? (*Hint:* Remember that clocks at signs *A* and *B* would be synchronized with the clock at *O* in such a way that they would read the speed of a light signal traveling between them and *O* to be 1 s/s. Knowing this, knowing the distance between *O* and *A*, and knowing the time that light from event *A* reached *O*, can you *infer* when that event must have happened?)

FIGURE 2.11

Observing a high-speed train. (Event *A* is shown in this diagram.)

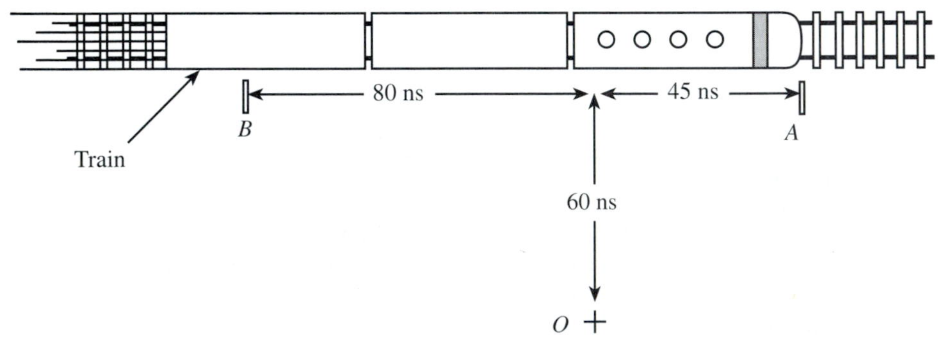

3

THE NATURE OF TIME

Philosophy is written in this great book (by which I mean the universe) which stands always open to our view, but it cannot be understood unless one first learns how to comprehend the language and interpret the symbols in which it is written, and its symbols are triangles, circles and other geometric figures, without which it is not humanly possible to comprehend even one word of it; without these, one wanders in a dark labyrinth.

Galileo Gallilei[1]

3.1 THREE KINDS OF TIME

A firecracker explodes. Some time later and somewhere else, another firecracker explodes. How much time passes between these two events? How can we measure that time interval?

In a newtonian universe, such questions would be easy to answer. We might measure the time between the events with a pair of synchronized clocks, one present at each event. We might instead measure the time between the events with a single clock that travels in such a manner that it arrives at each event's location just as it happens. Since all clocks in a newtonian universe register the same universal, absolute time, these methods (and any of a number of other valid methods) will yield the same result. It is unimportant what method is actually used.

[1]From *Il Saggiatore,* 1623. Quoted in C. W. Misner, K. S. Thorne, and J. A. Wheeler, *Gravitation,* San Francisco: Freeman, 1973, p. 304.

But in the last chapter we argued that in the real universe, universal and absolute time does not exist, and therefore the problem of measuring the time interval between two events is somewhat more problematic. In this chapter we will discover that there are three fundamentally different ways to measure the time interval between two events in the theory of special relativity and that *these different methods yield different results,* even if applied to the same two events!

Because it is important in the real universe to distinguish between the various methods used to measure the time interval between two events, we refer to the time interval determined by each method using a special technical name: we speak of the *coordinate time,* the *proper time,* and the *spacetime interval* between the events. The purpose of this chapter is to describe these distinct ways of measuring time and begin to uncover the relationships between them.

3.2 THE COORDINATE TIME BETWEEN TWO EVENTS

Once the clocks in an inertial reference frame have been satisfactorily synchronized, we can use them to measure the time coordinates of various events that occur in that frame. In particular, we can measure the time between two events A and B in our reference frame by subtracting the time read by the clock nearest event A when it happened from the time read by the clock nearest event B when it happened: $\Delta t_{BA} \equiv t_B - t_A$. Note that this method of measuring the time difference between two events requires the use of *two* synchronized clocks in an established inertial reference frame. Such a measurement therefore *cannot* be performed in the absence of an inertial frame.

In summary, the *coordinate time* between two events in a given frame is defined as follows:

Definition of Coordinate Time

The time measured between two events either by a *pair* of synchronized clocks at rest in a given inertial reference frame (one clock present at each event) or by a *single* clock at rest in that inertial frame (if both events happen to occur at that clock in that frame) is called the **coordinate time** between the events in that frame. The symbol Δt is used to represent the coordinate time between two events.

3.3 COORDINATE TIME IS FRAME-DEPENDENT

Imagine that the observer in some inertial reference frame (let us call this one the Other Frame: we will talk about a Home Frame in a bit) sets out to synchronize its clocks. In particular, let us focus on two clocks in that frame that lie on the x' axis and are equally separated from the master clock at $x' = 0$. At $t' = 0$, the observer causes the center clock to emit two flashes of light, one traveling in the $+x'$ direction and the other in the $-x'$ direction. Let the emission of these flashes from $x' = 0$ at $t' = 0$ be called the origin event O.

Now, since both the other clocks are the *same distance* from the center clock and since the speed of light is 1 (light-) second/second in every inertial reference frame, the left-hand clock will receive the left-going light flash (call the event of reception event A) at the *same* time that the right-hand clock receives the right-going flash (event B). By the definition of synchronization, both clocks should thus be set to read the *same* time at events A and B (a time in seconds equal to their common distance from the center clock).

This process is illustrated by the spacetime diagram in Fig. 3.1. Note that since all three clocks are at rest in this frame, their worldlines on the spacetime diagram are vertical. Moreover, since the speed of light is 1 s/s in this (and every other inertial) frame, the worldlines of the light flashes will have slopes of ± 1 on the spacetime diagram (i.e., they will make a 45° angle with each axis) as long as the axes have the same scale. On this diagram it is clear that events A and B really do occur at the same time in this "other" reference frame.

Now consider a second inertial reference frame (the Home Frame) within which the Other Frame is observed to move in the $+x$ direction at a speed β. Let us look at the same events from the vantage point of the Home Frame. For convenience, let us take the event of the emission of the flashes to be the origin event in this frame as well (i.e., event O is taken to occur at $t = x = 0$ in the Home Frame).

The observer in the Home Frame will agree that the right and left clocks in the Other Frame are always equidistant from the center clock in the Other Frame. Moreover, at $t = 0$, when the center clock passes the point $x = 0$ in the Home Frame as it emits its flashes, the right and left clocks are equidistant from the emission event. But during the time that the light flashes are moving to the outer clocks, the observer in the Home Frame observes the left clock to move up the x axis *toward* the flash coming toward it and the right clock to move up the x axis *away* from the flash coming toward it. As a result, the left-going light flash has *less* distance to travel to meet the left clock than the right-going flash does to meet the right clock. Since the speed of light is 1 in the Home Frame as well as in the Other Frame, this means that the left clock receives its flash first. *Therefore, event A is observed to occur* before *event B in the Home Frame.*

FIGURE 3.1
The synchronization of two clocks equally spaced from a center clock, as observed in the Other Frame. If the right and left clocks are set to agree at events A and B, they will be synchronized with each other.

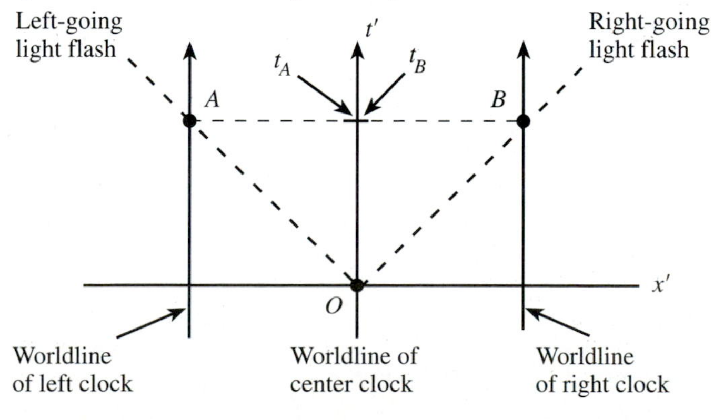

A spacetime diagram of the process as observed in the Home Frame is shown in Fig. 3.2. In drawing the worldlines of the clocks in question, it is important to note that the clocks are *not* at rest in the Home Frame. Their worldlines on a Home Frame spacetime diagram will be equally spaced lines with slopes of $1/\beta$, indicating that the clocks are moving to the right at a speed β. The light flashes have a speed of 1 s/s in the Home Frame (and every other frame), so their worldlines are drawn with a slope of ± 1 on the spacetime diagram.

In summary, the coordinate time between events A and B as measured in the Other Frame is $\Delta t' = 0$ (by construction in this case), but the coordinate time between these events as measured in the Home Frame is $\Delta t \neq 0$. We see that the coordinate times between the *same two events* measured in different reference frames are *not* generally equal. Thus coordinate time differences are said to be **relative** (i.e., they depend on one's choice of inertial reference frame).

Why? If each observer synchronizes the clocks in his or her own reference frame according to our definition, *each will conclude that the clocks in the other's frame are not synchronized.* Notice that the Other Frame observer has set the right and left clocks to read the same time at events A and B. Yet these events do *not* occur at the same time in the Home Frame. Therefore the observer in the Home Frame will claim that the clocks in the Other Frame are not synchronized. (Of course, the observer in the Other Frame will make the same claim about the clocks in the Home Frame.) The definition of synchronization that we are using makes perfect sense *within* any inertial reference frame, but it does not allow us to synchronize clocks in *different* inertial frames. In fact, the definition *requires* that observers in different inertial frames measure different time intervals between the same two events, as we have just seen.

In general, two observers in different frames will also disagree about the spatial coordinate separation between the events. Consider two events C and D that both occur at the center clock in the Other Frame, but at different times. Since the center clock defines the location $x' = 0$ in the Other Frame, the events will be measured to

FIGURE 3.2
The same events as observed in the Home Frame. In this frame, event B is measured to occur *after* event A.

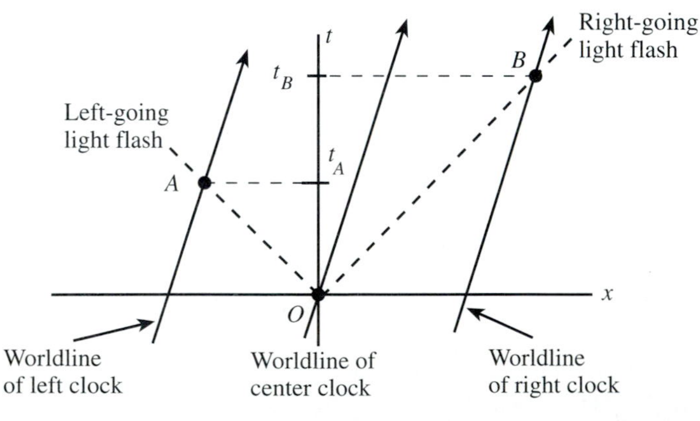

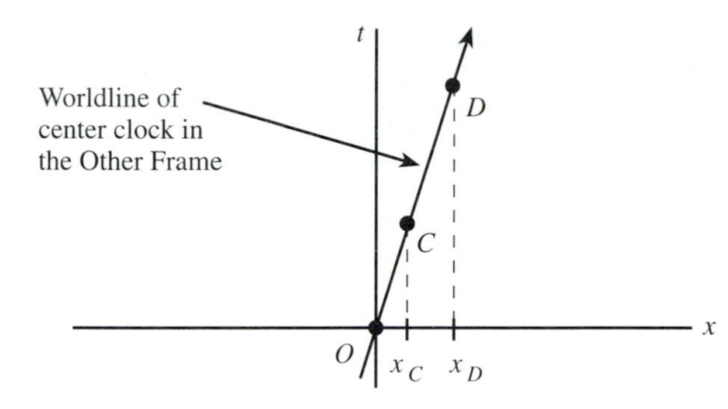

FIGURE 3.3
Events C and D both occur at the center clock in the Other Frame; therefore $\Delta x' = 0$ in that frame. But the center clock is measured to move with respect to the Home Frame between the events; therefore $\Delta x \neq 0$ in the Home Frame. (The spacetime diagram shows events as they would be plotted by the Home Frame observer.)

have the same x' coordinate in that frame, implying that $\Delta x' = 0$. But in the Home Frame, the center clock is measured to move in the time between the events, and so the two events do not occur at the same place: $\Delta x \neq 0$ (see Fig. 3.3).[1]

The point here is that the *coordinate differences between two events are frame-dependent quantities,* whether they are time coordinate differences or space coordinate differences: they will in general be measured to have different values in different inertial reference frames.

3.4 THE RADAR METHOD YIELDS THE SAME RESULTS

The basic reason why observers in different inertial frames disagree about whether the clocks in a given frame are synchronized or not is that synchronization is *defined* so that light flashes are measured to have a speed of 1 in every inertial frame: the frame dependence of coordinate time differences is a logical consequence of this assertion. This can be illustrated by considering the radar method of assigning spacetime coordinates; though the radar method does *not* involve the use of synchronized clocks, it does depend on the assumption that the speed of light is the same in every inertial frame. Does the radar method also imply that the coordinate time difference between two events is frame-dependent?

Figure 3.4 shows that it does. Figure 3.4*a* shows the observer in the Other Frame using the radar method to determine the space and time coordinates of event C. The observer in that frame will conclude that events C and D occur at the same time, where D is the event of the master clock at $x' = 0$ reading $t'_D = \frac{1}{2}(t'_B - t'_A)$, that is, the instant of time halfway between the emission of the radar pulse at t'_A and its reception at t'_B. According to the radar method, then, the coordinate time interval between events C and D is $\Delta t' = 0$. [Note: Radar and visible light are both examples of electromagnetic waves: they just have different frequencies. Both types of waves will thus move at a speed of 1 (light-)second / second = 1.]

[1]The frame dependence of the distance between events has nothing to do with clock synchronization or special relativity: this would be true even if time were universal and absolute, as assumed in galilean relativity.

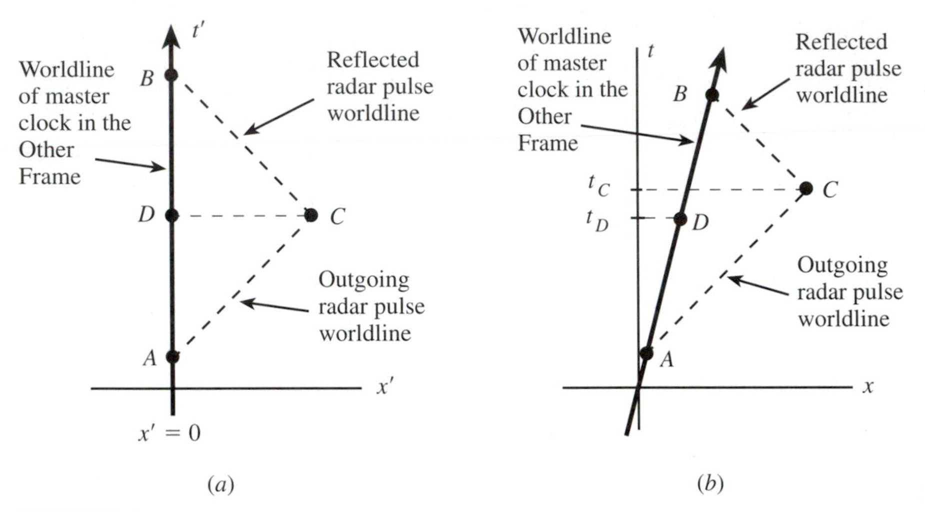

FIGURE 3.4
(a) In the Other Frame, events *C* and *D* are simultaneous, where *D* occurs at the master clock at a time halfway between the emission event *A* and the reception event *B*. The coordinate time difference between events *C* and *D* in this frame is thus $\Delta t' = 0$, according to the radar method. *(b)* In the Home Frame, the Other Frame's master clock moves to the right as time passes, so its worldline is slanted in a spacetime diagram based on measurements in that frame. On the other hand, radar pulse worldlines still have slope ± 1, as shown. This means that an observer in the Home Frame will conclude that event *C* happens after event *D*; the coordinate-time difference between the events is $\Delta t \neq 0$.

When the same sequence of events is viewed from the Home Frame, though, a different conclusion emerges (see Fig. 3.4*b*). According to observers in the Home Frame, the Other Frame's master clock is moving along the x axis with some speed β, so in a spacetime diagram based on measurements taken in the Home Frame, the worldline of that clock will appear as a slanted line (slope $= 1/\beta$) instead of being vertical. Radar pulse worldlines, on the other hand, still have slopes of ± 1, just as they did in the Other Frame spacetime diagram. The inevitable result (as you can see from the diagram) is that observers in the Home Frame are forced to conclude that event *C* occurs *after* event *D* does and, therefore, that the time difference between events *C* and *D* in the Home Frame is $\Delta t \neq 0$.

The point is that the relativity of coordinate time differences is a direct consequence of the fact that we are *defining* coordinate time by assuming that the speed of light is 1 in every inertial reference frame. Remember, though, that we *must* make this assumption if the laws of electromagnetism are to be consistent with the principle of relativity!

3.5 A GEOMETRIC ANALOGY[1]

It may be troubling that coordinate differences between events are not absolute but are instead frame-dependent. This is particularly true of the time coordinate separation: it

[1]This analogy is beautifully presented as a "parable" at the beginning of E. F. Taylor and J. A. Wheeler's *Spacetime Physics,* San Francisco: Freeman, 1966. The material presented in this section has been influenced by that presentation.

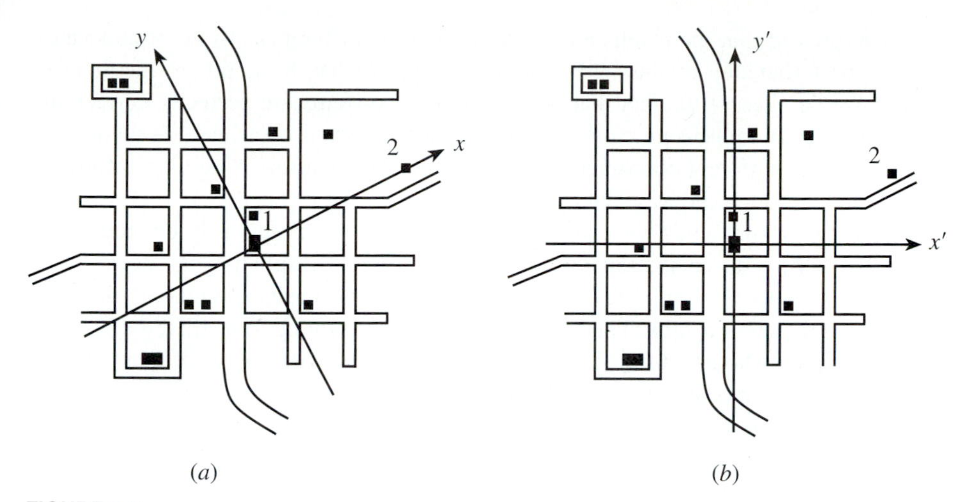

FIGURE 3.5
(a) Standard cartesian coordinate system superimposed on the town of Askew. *(b)* A more convenient coordinate system oriented 28° clockwise. On both maps, 1 is City Hall and 2 is the Statue of the Unknown Physicist.

is not easy to let go of the newtonian notion of absolute time! But the fact is that *we have no trouble at all with these ideas when they appear in a related but more familiar guise.*

Consider the hypothetical town of Askew, North Dakota. Most towns in rural America have streets that run north-south or east-west, but Askew has a problem. The surveyor who laid out the streets of this town in 1882 *tried* to calibrate his compass against the North Star the night before but in fact had forgotten exactly where the North Star was (it was a long time since he had this stuff in high school after all), and ended up choosing a star that turned out to be 28° east of the true North Star. So all the streets of Askew are twisted 28° from the standard directions.

Now, if we would like to assign x and y coordinates to points of interest in this town (or any town), we have to set up a cartesian coordinate system. It is *conventional* to orient coordinate axes on a plot of land so that the y axis points north and the x axis points east (see Fig. 3.5*a*). This is usually convenient as well, since the streets will be parallel to the coordinate axes. But there is no reason why this *has* to be done, and in Askew's case, it is actually more convenient to use a coordinate system oriented 28° to the east (Fig. 3.5*b*). (Note that the origin of both coordinate systems is chosen to be City Hall for the sake of simplicity.)

We can use any coordinate system that we choose to quantify the positions of points of interest in the town: coordinate systems are arbitrary human artifacts that we impose for our convenience on the physical reality of the town. But the coordinates we actually get for various points certainly do depend on the coordinate system used. For example, the coordinate differences between City Hall and the Statue of the Unknown Physicist in Memorial Park might be $\Delta y = 0$, $\Delta x = 852.0$ m in the standard coordinate system, whereas in the "convenient" coordinate system, they might be $\Delta y' = 399.9$ m, $\Delta x' = 752.3$ m.

Is it surprising that the results are different? Do the differences in the results cause us to suspect that one or the other coordinate system has been set up incorrectly? Hardly! We can accept the fact that both coordinate systems are perfectly correct and legal. We already *know and expect* that differently oriented coordinate systems on a plane will yield different coordinate measurements. This causes us no discomfort: we understand that this is the way that things are.

In an entirely analogous way, we have carefully and unambiguously defined a procedure for setting up an inertial reference frame and synchronizing its clocks. This definition happens to imply that spacetime coordinate measurements in different frames yield different results. This should be no more troubling to us than the fact that Askew residents who use different sets of coordinate axes will assign different coordinates to various points in town. *Coordinates have meaning only "relative" to the coordinate system or inertial frame being used.*

The only reason that the relativity of time coordinate differences is a difficult idea is that we do not have *common experience* with inertial reference frames moving with high enough relative speeds to display the difference. The kinds of inertial frames that we experience in daily life have relative speeds below 300 m/s, or about one-millionth the speed of light. (If for some reason we could only construct cartesian coordinate systems on the surface of the earth that differed in orientation by no more than a millionth of a radian, then we might think of cartesian coordinate differences as being "universal and absolute" as well!)

So, to summarize, the coordinate differences between points on a plane (or events in spacetime) are "relative" because coordinate systems (or inertial reference frames) are human artifacts that we *impose* on the land (or spacetime) to help us quantify that physical reality by assigning coordinate numbers to points on the plane (or events in spacetime). Because we are free to set up coordinate systems (or reference frames) in different ways, the coordinate differences between two points (or events) reflect not only something about their *real* physical separation but also something about the artificial choice of coordinate system (or reference frame) that we have made.

So, is it true then that *everything* is relative? Is there nothing that we can measure about the physical separation of the points (or events) that is "absolute," i.e., independent of reference frame?

There *is* in fact a coordinate independent quantity that describes the separation of two points on a plot of land: the *distance* between those two points. For example the distance between City Hall and the Statue of the Unknown Physicist in Askew, North Dakota, is $\Delta d = \sqrt{\Delta x^2 + \Delta y^2} = \sqrt{(852.0\text{ m})^2 + 0} = 852.0\text{ m}$ in the standard coordinate system and $\Delta d' = \sqrt{(\Delta x')^2 + (\Delta y')^2} = \sqrt{(399.9\text{ m})^2 + (752.3\text{ m})^2} = 852.0\text{ m}$ in the convenient coordinate system. It does not matter what coordinate system you use to calculate the distance: you *always* will get the same answer.

The distance between two points on a plot of ground thus reflects something that is deeply real about the nature of the plot of ground itself, without reference to the human coordinate systems imposed on it. This independence from coordinate systems arises because there is in fact a method of determining the distance between two points *without* using a coordinate system: lay a tape measure between the points! No coordinate system is required to do this. And since this method yields a certain definite result for the distance, calculations of this distance in *any* coordinate system should yield the same result if they are valid.

Of course, there are many ways that one could lay a tape measure between City Hall and the Statue of the Unknown Physicist. One could lay the tape measure along a straight path between the two points: this would measure the distance "as the crow flies," which is what is usually meant by the phrase "the distance between two points." But one might also lay the tape measure along other paths between the two points. One might, for example, lay the tape measure two blocks down Elm Street from City Hall, then one block over along Grove Avenue, then up Maple Street, and so on. This would measure a different kind of distance between the two points that we might call a *pathlength.*

Both the straight-line distance and the more general pathlength between two points can be measured directly with a tape measure and thus are quantities independent of any coordinate system. But the distance and the pathlength between two points may not be the same. In general, the pathlength between two points will depend on the path chosen and will always be greater than (or at best equal to) the straight-line distance.

To summarize, we have at our disposal three totally different ways to quantify the separation of two points on a plane. We can measure the *coordinate separations* between the points using a coordinate system. (The results will depend on one's choice of coordinate system.) We can measure the *pathlength* between the points with a tape measure laid along a specified path. (The result here will depend on the path chosen but is independent of coordinate system.) Or we can measure the *distance* between the points with a tape measure laid along a *special* path: a straight line. Because in this last case the path of the tape is uniquely specified, the distance between two points is a unique number that quantifies in a very basic way the separation of the points in space.

3.6 THE PROPER TIME AND THE SPACETIME INTERVAL

Consider two events. Label them A and B. Is there any way that we can measure the time between events A and B *without* using a reference frame, analogous to the way that we can measure the pathlength between two points on the plane without using a coordinate system?

We can avoid the use of a reference frame lattice *if we measure the time interval between these events with a clock that is present at both events.* In a manner analogous to laying a tape measure between two points so that it passes close by each point, we send a clock between the events along just the right path so that it is very close to each event as it occurs (see Fig. 3.6).

A tape measure stretched between two points marks off the distance between those points and presents a scale that can be laid right next to the two points for easy and unambiguous reading. In an entirely similar manner, a clock that travels between two events marks off the time between those events and presents its face at each event for easy and unambiguous reading. Since the clock's face is right there at each event, *everyone* looking at the clock will agree as to its reading as each event happens. The quantity measured by this clock is therefore frame-independent: it is measuring something basic about the *absolute* physical relationship between the events.

FIGURE 3.6
Measuring the time between two events with a clock that is present at both events (events A and B here are represented as firecracker explosions).

Definition of Proper Time

The time between two events measured by any clock present at both events is called a **proper time**[1] between those events. We will use the symbol $\Delta\tau$ to represent a proper time between two events. A proper time measured by a given clock is an absolute quantity independent of reference frame.

But there is one thing that the proper time between two events *might* conceivably depend on other than the events themselves: it might depend on the *worldline* that the clock takes in traveling from one event to the other, just as a pathlength measured by a tape measure depends on the path along which it is laid (see Fig. 3.7). We will see in the next chapter that the path dependence of proper time is a straightforward consequence of the principle of relativity; for now it is enough to understand that this path dependence is a *possibility* suggested by the geometric analogy.

In an experiment performed in 1971 by J. C. Hafele and R. E. Keating,[2] a pair of very accurate atomic clocks were synchronized, and then one was put on a jet plane and sent around the world while the other remained in the laboratory. These clocks were *both* present at the same two events (the event of the departure of the plane clock from the laboratory and the event of its return); both thus measure proper times between these events. But the worldlines followed by each of these clocks were very different; one clock's worldline was simply a straight line at constant position (in the reference frame of the earth's surface), while the other's worldline went around the world. When these initially synchronized clocks were again brought together, it was

[1]*Proper* is a misleading adjective here; in English this word has come to mean "appropriate," or "correct in morals or manners," when the meaning here is more accurately "proprietary," i.e., the time between the events measured *specially* by the clock in question. Perhaps *path time* would be a more appropriate term.

[2]*Science,* **117,** 168 July 14, 1972.

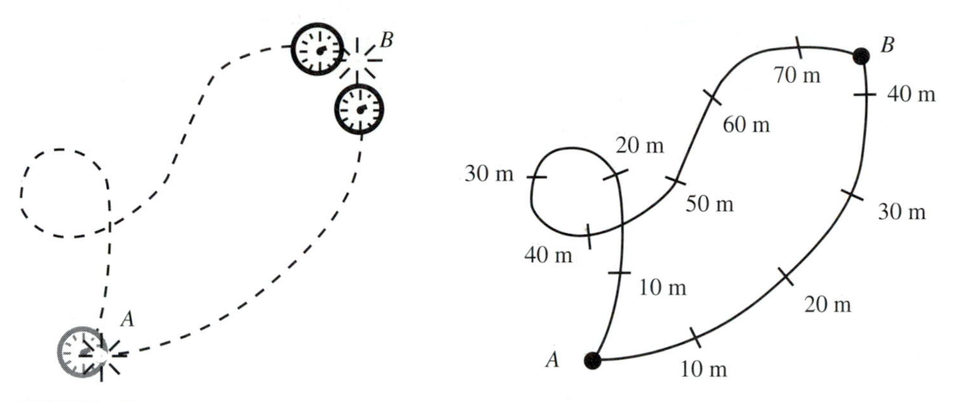

FIGURE 3.7
The proper time measured between two events (diagram at left) *may* depend on exactly how the clock travels between those events, just as the pathlength between two points (right) depends on the path along which it is measured.

found that they disagreed by several hundred nanoseconds (a difference that is in fact consistent with the *quantitative* prediction based on the theories of special and general relativity). The point is that it is not merely *possible* that the time that a clock measures depends on the nature of its worldline, it is an established experimental *fact*.

When measuring distances on the plane, we distinguish between the *pathlength* between two points measured along a certain path and the *distance* between the points: the *distance* is measured along the special path that is the straight line between the two points. Because the straight-line path is unique, the distance between two points along a straight line is a unique number that reflects something definite and unambiguous about the separation of those points in space.

Similarly, an inertial clock (a clock whose attached first-law detector registers no violation of Newton's first law) follows a unique and well-defined worldline through spacetime between two events. Observed in an inertial reference frame, a clock would travel between the events in a straight line at a constant velocity. Such a worldline defines a unique path in space, and since there is only one value of a constant velocity that will be just right to get the clock from one event to the other, the clock's velocity along that worldline is also uniquely specified.

Definition of Spacetime Interval

The **spacetime interval** between two events is defined to be the proper time measured by an *inertial* clock that is present at both events. This quantity is a unique, frame-independent number that depends on the separation of the events in space and time and *nothing else*. The symbol Δs is conventionally used to represent the spacetime interval between two events.

Now, it is important to note that the definitions of *coordinate time, proper time,* and the *spacetime interval* between two events overlap in certain special cases. The

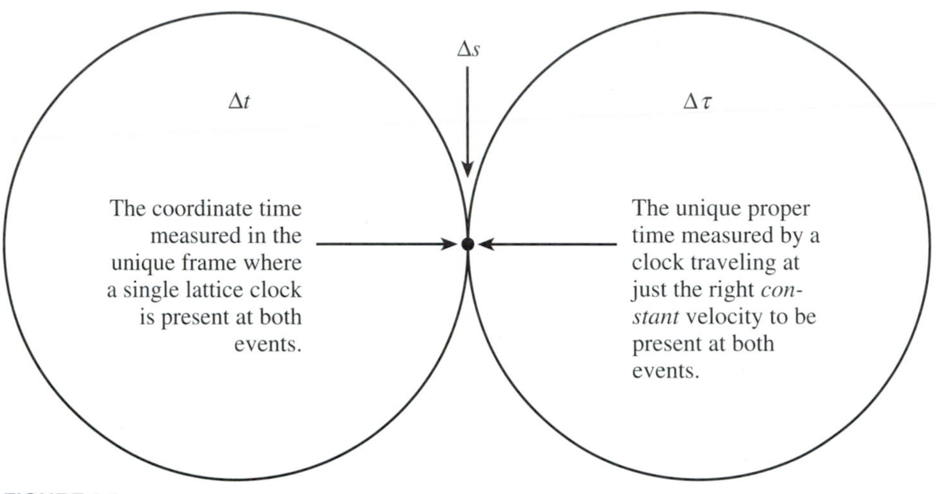

FIGURE 3.8
Let various points in the left circle represent the various coordinate times that observers in inertial frames moving at various different velocities might measure between two events. Let various points in the right circle represent the various proper times measured between the events by clocks present at both events but moving between the events along various different worldlines. The single point in common between these sets of possible time intervals between the events is the spacetime interval Δs between the events.

definition above makes it clear that the spacetime interval between two events is a *special case* of a proper time between two events, just as the distance between two points is a special case of the more general pathlength between two points. An inertial clock present at both events also measures the coordinate time between those events in the clock's own reference frame, since the time interval measured between two events by a clock or clocks at rest in any inertial reference frame is a coordinate time by definition. So the spacetime interval between two events is a special case of a proper time *and* a special case of a coordinate time (see Fig. 3.8).

3.7 SUMMARY: THE THREE KINDS OF TIME

In this chapter, we have seen that there are three fundamentally different ways of measuring the time between two events. We can measure the *coordinate time* Δt between the events in a given inertial frame with either a single clock (if the events happen to occur in the same place in that frame) but more generally a *pair* of clocks, one present at each event, that are at rest and have been synchronized in that frame. We can measure the *proper time* $\Delta\tau$ between the events using a single clock that is present at both events, and we can measure the *spacetime interval* Δs between the events using a single *inertial* clock present at both events.

The spacetime interval is actually a *special case* of a proper time (since the clock measuring the spacetime interval is present at both events) and *a special case* of coordinate time (since the time interval measured between the events is the coordinate time in the clock's own inertial frame).

The coordinate time measured between two events is a *frame-dependent* quantity. This is because measuring the coordinate time in general involves comparing times read from two synchronized clocks in a given frame, and as we have seen, clocks synchronized in one frame appear to observers in a different frame to be out of synchronization. Therefore, observers in different reference frames will inevitably disagree about what the coordinate time between two events really is.

On the other hand, the proper time (read *proprietary* time) $\Delta\tau$ measured by a given clock present at both events is a *frame-independent* quantity. This is because all observers in all reference frames can see for themselves what that clock registered at the first event and what it registered at the second event. Since no comparison of "synchronized" clocks is involved, there are no grounds for disagreement about the value of the time interval. The value of the proper time measured by different clocks may depend on the particular worldlines that they follow in going from one event to another, but the proper time measured by a given clock has a well-defined value everyone can agree on.

The spacetime interval Δs between two events is the proper time measured by an inertial clock present at both events (or alternatively, the coordinate time measured in an inertial frame where the events occur at the same place). Since for a given pair of events there is one unique worldline that carries the clock at constant velocity from one event to the other, the spacetime interval between two events has a *unique, frame-independent* value (*all* observers will agree that the clock measuring the spacetime interval between two events is in fact inertial and was in fact present at the events and can read for themselves the value registered by the clock at each event, and so all agree on the unique value of the spacetime interval). The spacetime interval thus characterizes something *absolute* about the separation of these events in spacetime, in distinction to the *relative* (frame-dependent) value of the coordinate time between the events and in distinction to the worldline-dependent value of the proper time between the events.

We have also been developing in this chapter a geometric *analogy* between spacetime and the geometry of a two-dimensional plane. Points on a plane correspond to events in spacetime, paths on a plane correspond to worldlines in spacetime, spacetime diagrams of events in spacetime correspond to maps of points on a plane, and so on. The most important part of the analogy for us at present is that the three possible ways of measuring the time interval between two events discussed in this chapter (*coordinate time, proper time,* and the *spacetime interval*) are analogous to the three possible ways of characterizing the separation of two points on a plane (*coordinate separation, pathlength,* and *distance,* respectively).

Tables 3.1 and 3.2 organize and summarize the ideas presented in this chapter.

In the next chapter, we will continue our investigation of the nature of time by using the metric equation to derive an equation that links coordinate time and the proper time measured by a clock traveling along a general curved worldline. This will complete our mathematical description of the relationship between the three kinds of time.

TABLE 3.1
THE THREE KINDS OF TIME

	Coordinate time	Proper time	Spacetime interval
Definition	The time between two events measured in an inertial frame by a *pair of synchronized clocks,* one present at one event, the other present at the other event. (If both events happens to occur at the same place, a single clock suffices.)*	The time between two events as measured by a *single clock that is present at both events.* (Its value depends on the worldline that the clock follows in getting from one event to the other.)	The time between two events as measured by an *inertial* clock that is present at both events. (Because an inertial clock follows a unique worldline between the events, the spacetime interval is unique.)
Conventional symbol	Δt	$\Delta \tau$	Δs
Is value frame-independent?	No	Yes	Yes
Geometric analogy	Spatial coordinate differences	Pathlength	Distance

**Note:* Alternatively, the coordinate time difference between two events might be inferred from measurements of the spacetime coordinates of those events using the radar method.

TABLE 3.2
THE GEOMETRIC ANALOGY

Plane geometry	Spacetime geometry
Map	Spacetime diagram
Points	Events
Paths or curves	Worldlines
Coordinate systems	Inertial reference frames
Relative rotation of coordinate systems	Relative velocity of inertial reference frames
Differences between coordinate values	Differences between coordinate values
Pathlength along a path	Proper time along a worldline
Distance between two points	Spacetime interval between two events

PROBLEMS

3.1 THE THREE KINDS OF TIME. Imagine two clocks P and Q. Both clocks leave the spatial origin of an inertial frame S at time $t = 0$: call this the origin event O. Both clocks move along the $+x$ axis, with clock P originally traveling at a speed of about 4/5, while Q travels at a speed of about 1/5. After a while, however, clock P decelerates, comes to rest, and then begins to move back toward the origin. A short time later, clock P collides with the slower clock Q, which has been moving with constant speed up the x axis during all this. Let the collision of the clocks be event A.

a Draw a *qualitatively* accurate spacetime diagram of the situation described above, showing the worldlines of clocks P and Q and the locations of events O and A.

b Assume that clocks P and Q were both synchronized with the clock at the origin of frame S when they left the origin. Will P and Q necessarily read the same value when they collide? Explain.

c An observer in frame S measures the time between events O and A with a pair of synchronized clocks (one at the spatial location of event O and one at the spatial location of event A). Clocks P and Q each also register a time between these events. Which clock(s) measure proper time between the events? The spacetime interval between the events? The coordinate time between the events?

3.2 THE THREE KINDS OF TIME. Clock P is at rest alongside a racetrack. A jockey on horseback checks his watch against clock P as he passes it during the first lap (call this event A) and then checks his watch again as he passes clock P the second time (call this event B). Which clock (clock P or the watch) measures the spacetime interval between events A and B? Which measures proper time? (Careful!) Do either of the clocks measure coordinate time between the events in the earth frame of reference? Discuss.

3.3 THE THREE KINDS OF TIME. Alyssa is a passenger on a train moving at a constant velocity. Alyssa synchronizes her watch with the station clock as she passes through the Banning town station and then compares her watch with the station clock as she passes through the Centerville town station farther down the line. Is the time that she measures between the events of passing through these towns a proper time? Is it a coordinate time? Is it the spacetime interval between the events? If one subtracts the Centerville station clock reading from the Banning station clock reading, what kind of time interval between the events does one obtain? Defend your answers carefully. (Treat the ground as an inertial frame.)

3.4 OUT OF SYNCH. Two firecrackers A and B are placed at $x' = 0$ and $x' = 100$ ns, respectively, on a train moving in the $+x$ direction relative to the ground frame. According to synchronized clocks on the train, both firecrackers explode simultaneously. Which firecracker explodes first according to synchronized clocks on the ground? Explain carefully. (*Hint:* Study Fig. 3.2.)

3.5 TV TIME. Imagine that in the year 2065, you are watching a live broadcast from the space station at the planet Neptune, which is 4.0 light-hours from Earth at the time. (Assume that the TV signal from Neptune is sent to Earth via a laser light communication system.) At exactly 6:17 P.M. (as registered by the clock on your desk) you see a technician on the TV screen suddenly exclaim, "Hey! We've just detected an alien spacecraft passing by here." Let this be event A. Exactly 1 h later, the alien spaceship is detected passing by Earth: let this be event B. Assume that the Earth and Neptune stations can be considered parts of the inertial reference frame of the solar system, and assume that the spaceship travels at a constant velocity.

a On your TV screen, you can see the face of a clock sitting on the technician's desk. What time should you see on this clock face if that clock is synchronized with yours?

b What is the coordinate time between events A and B in the solar system frame? Defend your response carefully.

c What is the speed of the alien spaceship as measured in the solar system frame?

d What kind(s) of time would the spaceship's clock measure between events A and B?

3.6 THE THREE KINDS OF TIME. A particle accelerator is a device that boosts subatomic particles to speeds close to that of light. Such an accelerator is typically shaped like a ring (which may be several kilometers in diameter); the particles are constrained by magnetic fields to travel inside the ring. Imagine such an accelerator having a radius of 2.998 km. Assume that there are two synchronized clocks (P and Q) located on opposite sides of the ring. A certain particle in the ring is measured to travel from clock P to clock Q in 34.9 μs as registered by those clocks. Let event A be the particle's departure from clock P and event B be the particle's arrival at clock Q. Assume that the particle contains an internal clock that also measures the time between these events.

a What is the speed of the particle in the laboratory frame?

b Does the synchronized pair of laboratory clocks measure the proper time, the coordinate time, or the spacetime interval between events A and B?

c Does the particle's internal clock measure the proper time, the coordinate time, or the spacetime interval between events A and B?

3.7 THE SYNCHRONIZATION DISCREPANCY. (Challenging!) Consider Fig. 3.2. The spacetime diagram shown was used to argue that clocks synchronized in the Other Frame will *not* be synchronized in the Home Frame. Imagine that the spatial separation between each side clock and the center clock is $L = 12$ ns as measured in the Home Frame and that the speed of the clocks relative to the Home Frame is $\beta = 0.40$ in SR units. Find the time separation $t_A - t_B$ between events A and B as measured in the Home Frame. [*Hint:* This is a *tricky* problem, but it is not as impossible as it looks. Consider the left clock. In the time between $t = 0$ and $t = t_A$, this clock moves a distance βt_A toward the light flash coming toward it. Thus the total distance that the light flash has to cover in this time interval is $L - \beta t_A$. But since the light flash travels with unit speed in the Home Frame (and every frame), the *time* it takes to travel to the clock is equal to the *distance* it has to travel (in SR units). Write this last sentence as an equation, and then solve the equation for t_A. Repeat for the right-hand clock.]

4

THE METRIC EQUATION

The most incomprehensible thing about the world is that it is comprehensible.

Albert Einstein[1]

4.1 INTRODUCTION

In the last chapter, we discussed three different ways to measure the time interval between two events. An observer in an inertial reference frame can measure the *coordinate time* Δt between the events by comparing the reading of the clock present at one event with the synchronized clock present at the other event. Since observers in different reference frames will disagree about whether a given pair of clocks is synchronized, the coordinate time Δt measured between the events depends on the reference frame used. One can avoid this problem by measuring the time between the events by a *single* clock that moves between the events so that it is present at both. Such a clock measures a proper (proprietary) time $\Delta\tau$ between the events. The value of such a proper time, while frame-independent (in the sense that *all* observers will agree on what a given clock present at both events actually registers between those events), is known experimentally to depend on the worldline traveled by the clock as it moves from one event to the other. There is, however, one and *only* one worldline that takes a clock from one event to the other at a *constant velocity.* The proper time between the events measured by an *inertial* clock is thus a unique, frame-independent number that

[1]Quoted in A. P. French (ed.), *Einstein: A Centenary Volume,* Cambridge, Mass.: Harvard, 1979, p. 53.

depends on the spacetime separation of the two events and nothing else. This unique proper time is called the *spacetime interval* Δs.

In the last chapter, we also discussed an analogy that compared these different ways of measuring the time between events in spacetime with different ways of measuring the displacement of two points on a plane. The coordinate time corresponds to the north-south (or east-west) *coordinate displacement* between the points. Since surveyors using different coordinate systems will disagree about the exact direction of "north," the value of the north-south displacement between two points will depend on one's choice of coordinate system. The proper time corresponds to the *pathlength* between the two points measured along a certain path using a tape measure. Since measuring the pathlength does not require determining where north is, its value is independent of the coordinate system but does depend on the path chosen. The spacetime interval corresponds to the straight-line *distance* between the points. There is one and only one straight line between a given pair of points, so the pathlength measured along this line is a unique, coordinate-independent number that depends on the spatial separation of the points.

As discussed in Sec. 3.5, we can actually *calculate* the distance Δd between two points on the plane using the coordinate displacements Δx and Δy between the points (as measured in *any* given coordinate system) and the pythagorean theorem:

$$\Delta d^2 = \Delta x^2 + \Delta y^2 \tag{4.1}$$

The amazing thing about this formula is that while the values Δx and Δy between two points depend on one's choice of coordinate system, the distance Δd calculated from these does not.

It turns out that there is an analogous formula that links the coordinate time Δt and spatial displacements Δx, Δy, and Δz between two events measured in any given inertial reference frame with the frame-independent spacetime interval Δs between the events. This equation, called the **metric equation,** provides the crucial key needed to *escape* the "relativity" of inertial reference frames and quantify the separation of the events in *absolute* (frame-independent) terms.

Our purpose in this chapter is to *derive* the metric equation and describe some of its immediate consequences. In the next chapter we will show how the metric equation can be used to compute proper times along more general worldlines as well.

4.2 THE METRIC EQUATION

The derivation that follows is the very core of the special theory of relativity. The metric equation is the key to understanding *all* the unusual and interesting consequences of the theory of relativity. You should make a special effort to understand this argument thoroughly.

What we want to do is compare the time interval Δt between two events measured in some inertial frame (call it the Home Frame) with the time Δs measured by a clock moving at a constant velocity that is present at both events. To make this argument easier, I want to consider a special kind of clock called a *light clock.* An idealized light clock is shown in Fig. 4.1. It consists of two mirrors a fixed distance L apart and a

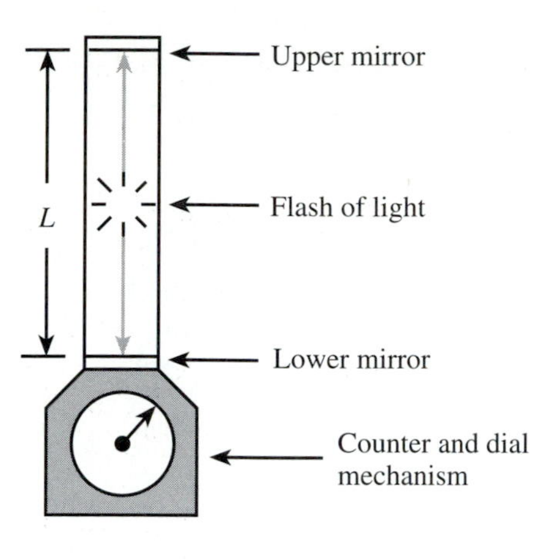

FIGURE 4.1
Schematic diagram of a light clock. Each "tick" of the light clock represents the passage of a time interval equal to $2L$ (if L is measured in seconds).

flash of light that bounces back and forth between the mirrors. Each time the flash of light bounces off the bottom mirror, a detector in that mirror sends a signal to an electronic counter. The clock dial thus essentially registers the number of round-trips that the light flash has completed. Since the speed of light is *defined* to be 1 second of distance per 1 second of time in *any* inertial frame, this clock should be calibrated to register a time interval of $2L$ (where L is expressed in seconds) per "tick" (i.e., each time the light flash bounces off the bottom mirror): the clock will then read correct time as long as it is inertial.

Now consider an arbitrarily chosen pair of events A and B. Let the coordinate time interval and spatial separation between these events (as measured in the Home Frame) be Δt and $\Delta d = \sqrt{\Delta x^2 + \Delta y^2 + \Delta z^2}$, respectively. Also imagine that we have a light clock moving between these events (with its beam path oriented perpendicular to its direction of motion) at just the right constant velocity to be present at both events. To simplify our argument, let us also imagine that the length L between the light-clock mirrors has just the right value so that events A and B happen to coincide with successive ticks of the light clock (in principle, we could always adjust L to make this true for the two events we want to examine).

In the inertial frame of the light clock, both events occur at the clock face, and the clock's light flash completes exactly one round-trip. The time interval recorded by this clock between events A and B is thus exactly $2L$. Since this clock is present at both events, it registers the *spacetime interval* between these events, so $\Delta s = 2L$.

In the Home Frame, the time of each event is registered by the clock nearest the event; since the events occur at different places, the coordinate time interval between the events will be determined by taking readings from a *pair* of clocks. In this frame, the light clock is observed to move the distance Δd in the time interval Δt . This means that the light flash will be observed in the Home Frame to follow the zigzag shown in Fig. 4.2.

The distance that the light flash travels in the Home Frame is (by the pythagorean theorem)

$$2\sqrt{L^2 + (\Delta d/2)^2} = \sqrt{4L^2 + \Delta d^2} = \sqrt{(2L)^2 + \Delta d^2} \tag{4.2}$$

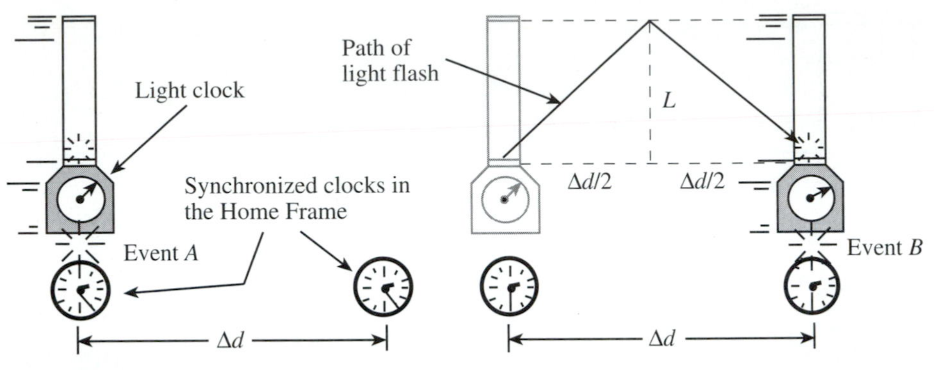

FIGURE 4.2
As the light clock moves from event *A* to event *B* in the Home Frame, its internal light flash will be observed to follow the zigzag path shown. The total distance covered by the flash in this frame is $2\sqrt{L^2 + (\Delta d/2)^2}$.

Since the synchronized clocks in the Home Frame must (by definition of synchronization) measure the speed of light to be 1, the coordinate time interval Δt registered on the pair of synchronized clocks in the Home Frame *must* be equal to the distance that the light flash traveled between the events:

$$\Delta t = \sqrt{(2L)^2 + \Delta d^2} \tag{4.3}$$

But we saw above that the light clock registers the spacetime interval between the two events to be $\Delta s = 2L$. Plugging this into Eq. (4.3) and squaring both sides, we get

$$\Delta t^2 = \Delta s^2 + \Delta d^2 \quad \text{or} \quad \Delta s^2 = \Delta t^2 - \Delta d^2 \tag{4.4}$$

Since $\Delta d^2 = \Delta x^2 + \Delta y^2 + \Delta z^2$ (where Δx, Δy, and Δz are the coordinate differences measured between the events in the Home Frame), we have, finally,

$$\boxed{\Delta s^2 = \Delta t^2 - \Delta x^2 - \Delta y^2 - \Delta z^2} \tag{4.5}$$

This extremely important equation links the *frame-independent* spacetime interval Δs between any two events to the frame-*dependent* coordinate separations Δt, Δx, Δy, Δz measured between those events in *any arbitrary inertial reference frame!* Note that we have not sacrificed anything by using a light clock instead of some other kind of clock in this argument: the speed of light is 1 in *any* inertial frame, and thus any decent clock that we construct must agree with what the light clock says. The only real limitation to our argument is that Δt must be greater than Δd for the two events in question so that it is *possible* for a light flash to travel between the events. (Note that if

$\Delta t < \Delta d$, Eq. (4.5) yields an imaginary number for Δs, an absurd result indicating that the conditions of the proof have been violated.[1])

Just as the spacetime interval Δs between two events is analogous to the distance Δd between two points on a plane, the formula $\Delta s^2 = \Delta t^2 - \Delta x^2 - \Delta y^2 - \Delta z^2$ is directly analogous to the pythagorean relation $\Delta d^2 = \Delta x^2 + \Delta y^2 + \Delta z^2$. Note that the pythagorean relation also relates a coordinate-independent quantity (the distance Δd between two points) with quantities whose values depend on the choice of coordinate system (the coordinate differences Δx, Δy, and Δz). Indeed, the formula for the spacetime interval would be just like a four-dimensional version of the pythagorean theorem if it were not for the minus signs that appear. We will see that these minus signs have a variety of interesting and unusual consequences.

Equation (4.5) is called the **metric equation.** It is the link between our human-constructed reference frames and the absolute physical reality of the separation between two events in space and time. It is difficult to overemphasize the importance of this equation: virtually all the rest of our study of the theory of relativity will revolve around the implications of this equation.

4.3 DISPLACEMENTS PERPENDICULAR TO THE LINE OF MOTION

The previous argument *assumes* that the vertical length L between the light-clock mirrors is the same in both the light-clock frame (where it was used to compute the spacetime interval) and the Home Frame (where it was used to compute the coordinate time). But how do we know that this is true? Since coordinate differences between events are generally frame-dependent, what gives us the right to assume that the displacement between the mirrors has the same value in both frames?

In this section, I will argue that if we have two reference frames in relative motion along a given line, any displacement measured *perpendicular* to that direction of motion will have the *same* value in both reference frames. I will demonstrate that this statement follows directly from the principle of relativity.

The proof presented here will be a proof by contradiction. We will assume that there *is* a contraction (or expansion) effect that applies to perpendicular lengths and show that the existence of such an effect contradicts the principle of relativity. Turned around, this argument will then imply that if the principle of relativity is true, no such effect can exist.

Consider two inertial reference frames (a Home Frame and an Other Frame) in standard orientation (see Sec. 1.9) so that the line of relative motion is along the frames' common x and x' axes. In each frame we set up a measuring stick along the y or y' direction with spray-paint nozzles set 1.00 m apart (as shown in Fig. 4.3). Note in the diagram that the common x and x' axes (which lie along the line of relative motion of the frames) are perpendicular to the plane of the paper. This means that as the frames move relative to each other, one of the measuring sticks will move *into* the

[1]Even so, the metric equation still yields a meaningful frame-independent value, as we will discuss in Chap. 7.

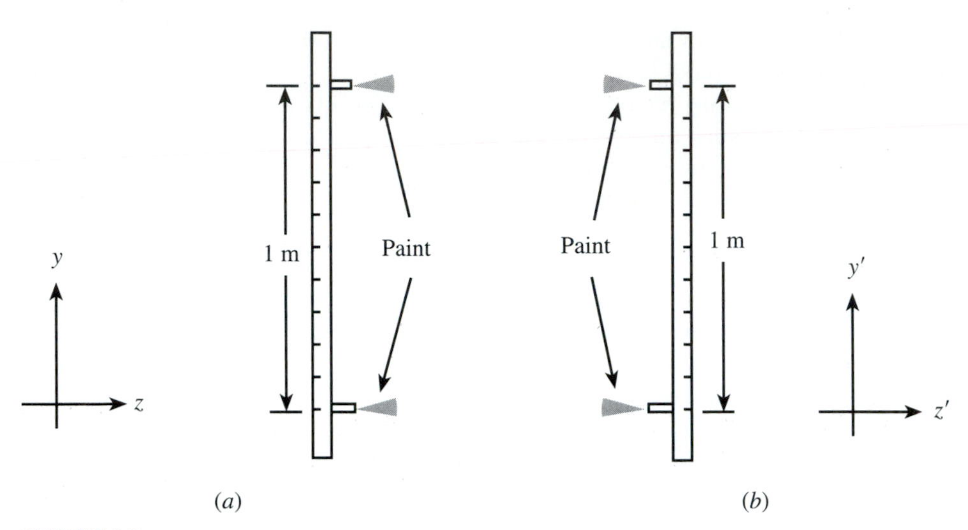

FIGURE 4.3
(a) The paint nozzles on the Home Frame measuring stick are exactly 1.00 m apart in the y direction, as measured in the Home Frame. They point at their counterparts in the Other Frame so that stripes will be painted on that frame's measuring stick as it moves by. The x axis points directly *into* the plane of the paper here. *(b)* The paint nozzles on the Other Frame measuring stick are exactly 1.00 m apart in the y' direction, as measured in the Other Frame. They point at their counterparts in the Home Frame so that stripes will be painted on that frame's measuring stick as it moves by. The x' axis points directly *into* the plane of the paper here.

paper and the other *out* of the paper. The paint nozzles in each frame are pointed in the direction of the secondary frame's measuring stick, so as the two measuring sticks pass each other, they will spray paint stripes on each other.

Now imagine that there exists some kind of contraction effect so that an observer in the Home Frame measures the measuring stick at rest in the Other Frame to be vertically contracted. This means that an observer in the Home Frame will measure the distance between the spray-paint nozzles on that stick to be *less* than 1.00 m apart. This in turn means that the stripes painted by these nozzles will be less than 1.00 m apart in the Home Frame: they will be painted *inside* the nozzles on the measuring stick at rest in the Home Frame. Conversely, the nozzles on the measuring stick at rest in Home Frame will paint stripes on the stick in the Other Frame which are *outside* the nozzles in that frame (see Fig. 4.4).

The principle of relativity requires that the laws of physics be exactly the same in any inertial reference frame. More specifically, this means that if you perform exactly the same experiment in two inertial reference frames, one should get exactly the same result. There should be *no* way of experimentally distinguishing the two frames.

Now, in the Other Frame, it is the Home Frame stick that is moving. Therefore, the principle of relativity requires that an observer in the Other Frame *must* measure the Home Frame stick to be contracted, just as the Home Frame observer measured the Other Frame stick to be contracted. This in turn means that the stripes painted by the Other Frame stick will be *outside* the Home Frame stick's nozzles and the stripes

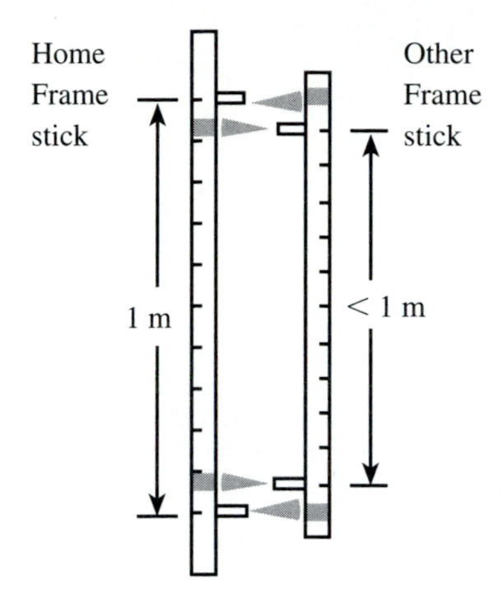

FIGURE 4.4
In the Home Frame, the Home Frame stick is at rest and the Other Frame stick is moving (directly *into* the paper). We are *assuming* that the moving stick is measured to be contracted in the Home Frame. This means that the Other Frame stick will paint marks on the Home Frame stick which are *inside* its nozzles and that the Home Frame stick will paint marks on the Other Frame stick which are *outside* its nozzles.

painted by the Home Frame stick will be *inside* the Other Frame stick's nozzles, as shown in Fig. 4.5.

Now compare Figs. 4.4 and 4.5: they describe a logical contradiction. In Fig. 4.4, stripes are being painted on the Home Frame stick *inside* its nozzles. In Fig. 4.5, stripes are being painted on the Home Frame stick *outside* its nozzles. These cannot be simultaneously true! The paint marks on the Home Frame stick are permanent and unambiguously visible to all observers in *every* reference frame. They *cannot* be "inside" the nozzles according to some observers and "outside" to others. So *either* Fig. 4.4 *or* Fig. 4.5 must be true, but *not both.* But the principle of relativity *requires* that *both* be true!

How can we resolve this conundrum? The only way is to reject the hypothesis that got us into this trouble in the first place, i.e., the hypothesis that distances measured *perpendicular* to the line of relative motion of the frames have different values in the

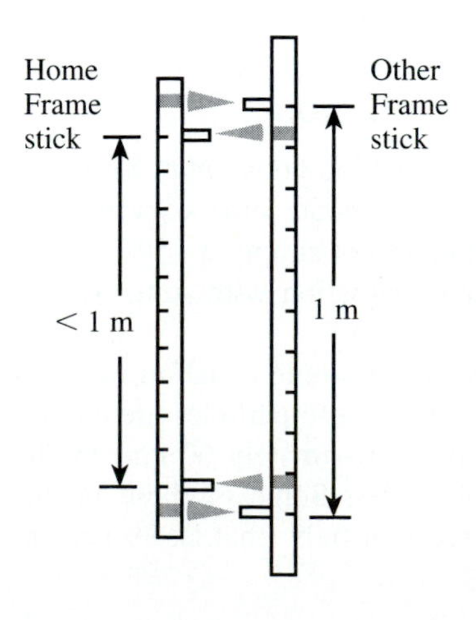

FIGURE 4.5
In the Other Frame, the Other Frame stick is at rest and the Home Frame stick is moving (directly *out* of the paper). We are assuming that the moving stick is measured to be contracted in the Other Frame. (This is required by the principle of relativity if the contraction exists in the Home Frame.) This means that the Home Frame stick will paint marks on the Other Frame stick which are *inside* its nozzles and that the Other Frame stick will paint marks on the Home Frame stick which are *outside* its nozzles.

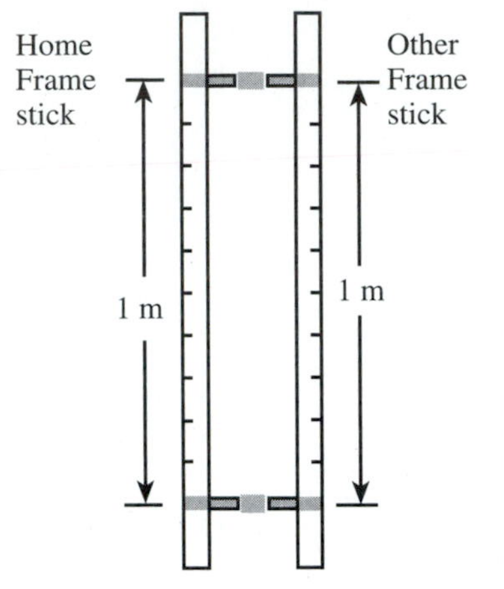

FIGURE 4.6
View in either frame. If we assume that there is *no* contraction-expansion effect, our problems disappear. Each stick paints across the nozzle of the other, and the situation is totally symmetric. Identical experiments in each frame then lead to identical results, as required by the principle of relativity. But this implies that observers in each frame will measure the distance between the other's spray-paint nozzles to have the same value.

two frames. If we assume that there is *no* contraction (or expansion) effects operating between the frames, then there is no problem with the principle of relativity: as shown in Fig. 4.6, both sticks will paint stripes across each other's nozzles. The situation is exactly the same in both frames, and the contradiction disappears.

This argument forces us to conclude:

Any displacement measured perpendicular to the line of relative motion of two inertial frames must have the same value in both frames.

(Note that the argument presented above *cannot* be applied to distances measured *parallel* to the x axis, as two sticks lying along the x axis will simply paint a long stripe down each other's length as they pass each other. No information about the length of either measuring stick can be extracted from this stripe.)

This means that the distance L between the mirrors used in the derivation of the metric equation does in fact have the same value in the light-clock frame as it does in the Home Frame, so our derivation is correct.

4.4 EXPERIMENTAL SUPPORT FOR THE METRIC EQUATION

Careful and compelling as the derivation of the metric equation above may be, we as physicists should not simply accept such an equation without some experimental confirmation: physics is a study of the physical world and not simply a collection of pretty logical arguments. We need to support the metric equation with some experimental evidence.

One of the classic experiments testing the validity of the metric equation involves the detection of certain subatomic particles called **muons.** These particles are continually generated in the upper atmosphere (at heights of approximately 60 km) by the interaction of cosmic rays with atmospheric gas molecules. Some of these muons stream downward toward the earth with speeds in excess of 0.99 (that is, 99 percent of c).

Now, muons are unstable particles decaying after a short period of time into lighter particles. The half-life of muons at rest in a laboratory is about 1.52 μs. This means that if you have N muons at a certain time, after 1.52 μs you will have $N/2$ left, after another 1.52 μs you will have $N/4$ left, and so on. You can imagine a muon to contain a tiny clock; there is a 50 percent chance that the muon will decay during each 1.52-μs time interval registered by this clock.

It is possible to build a muon detector that will count the number of muons reaching it from a particular direction and traveling at a particular speed. Imagine building two detectors that register muons traveling vertically downward at a speed of roughly 0.994 as measured in the earth's frame of reference. We place one such detector at the top of a mountain and count the number of muons that it sees per unit time, and another such detector at the foot of the mountain 1907 m (≈ 6.36 μs of distance) lower and count the number of muons that it sees per unit of time.

Let us follow a single muon that happens to go through both detectors. Let event A be this muon passing through the upper detector and event B this muon passing through the bottom detector. The distance Δd between these events in the earth's frame of reference is 6.36 μs in SR units. The coordinate time interval between these events measured in the earth's frame is simply the time required for a muon traveling at a speed of 0.994 to traverse this distance: $\Delta t = \Delta d/v = 6.36\ \mu\text{s}/0.994 = 6.40\ \mu\text{s}$.

Note that since the muon is present at each of these events by definition and moves between them at a constant velocity of 0.994 downward, the clock *inside* this muon measures the spacetime interval Δs between the events. If the newtonian conception of time were true, all clocks would measure the same time interval between two events, implying that $\Delta s = \Delta t = 6.40\ \mu\text{s}$. This corresponds to $6.40\ \mu\text{s}/1.52\ \mu\text{s} \approx 4.21$ muon half-lives, so most of the hypothetical muon's comoving siblings that make it through the top detector would in fact decay before reaching the bottom detector. Specifically, if N muons make it through the top detector, we would expect to see N times $(0.5)^{4.21} \approx N/18.5$ make it to the bottom detector if the newtonian assumption about time is true.

But if the metric equation is true, the spacetime interval between the events would actually be $\Delta s = (\Delta t^2 - \Delta d^2)^{1/2} \approx [(6.40\ \mu\text{s})^2 - (6.36\ \mu\text{s})^2]^{1/2} \approx 0.714\ \mu\text{s}$. This time, which is the time that our muon (and its comoving siblings) measures between the events, is only $0.714\ \mu\text{s}/1.52\ \mu\text{s} \approx 0.47$ of a muon half-life, meaning that most of the muons will *not* decay before they reach the bottom detector. Specifically, if N muons pass through the upper detector, N times $(0.5)^{0.47} \approx N/1.38$ should make it to the bottom detector.

In summary, the newtonian conception of time leads to the prediction that the ratio of the number of muons passing through the upper detector to the number passing through the lower detector should be 18.5, while the metric equation predicts that the same ratio should be 1.38. This is a substantial difference that can be easily measured.

This experiment was done in the early 1960s by D. H. Frisch and J. B. Smith.[1] They reported observing the ratio to be 1.38 (within their experimental error), thus confirming the metric equation (and utterly refuting the newtonian conception of time by a substantial margin).

[1] *Am. J. Phys.*, **31**: 342, 1963.

4.5 THE GEOMETRY OF SPACETIME IS NOT EUCLIDEAN

We have found the analogy between ordinary euclidean plane geometry and spacetime geometry to be very illuminating, and this basic analogy will remain fruitful as we continue to develop the consequences of the theory of relativity. Nevertheless, it is important at this point to describe some of the important *differences* between euclidean geometry and spacetime geometry that are a result of the negative signs in the metric equation $\Delta s^2 = \Delta t^2 - \Delta x^2 - \Delta y^2 - \Delta z^2$ that do *not* appear in the corresponding pythagorean theorem $\Delta d^2 = \Delta y^2 + \Delta x^2$.

One important difference concerns the representation of distances on a map and spacetime intervals on a spacetime diagram. If one prepares a scale drawing (e.g., a map) of various points on a plane, the distance between any two points on the map is *proportional* to the actual distance between those points in space. That is, distances on the drawing directly correspond to distances in the physical reality being represented. In Fig. 4.7*a*, for example, to determine the distance between City Hall and the Statue of the Unknown Physicist, one need merely measure the distance (in inches) between the two sites on the map shown above and multiply by the conversion factor 1000 m = 1 in. It does not matter how the line between the two sites is oriented or where the sites are located on the drawing: the distance in the physical space being represented by the map is always proportional to the distance measured on that map.

However, it is *not* true that the displacement between two points on a spacetime diagram is proportional to the spacetime interval between the corresponding events. This is illustrated in Fig. 4.7*b*.

FIGURE 4.7
(a) A scale drawing of Askew, North Dakota (1 in = 1 km). The actual distance between (1) City Hall and (2) the Statue of the Unknown Physicist is 852 m. The distance between City Hall and the bandshell at (3) Higgenbottom Park is also 852 m. The length of both arrows on the drawing is 0.852 inch. *(b)* Both event *B* and event *C* are the same distance from event *A* on the spacetime diagram, but the spacetime interval between *A* and *B* is 4 s, while the spacetime interval between *A* and *C* is zero (since $\Delta t = \Delta x$)!

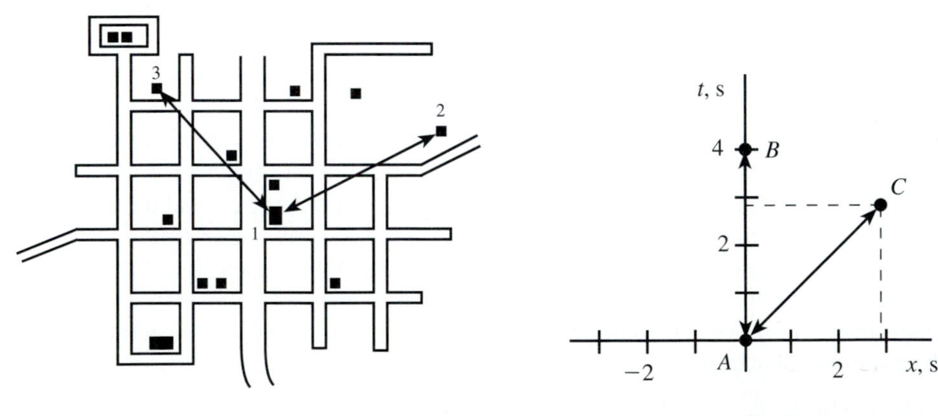

We see that a spacetime diagram accurately displays the spacetime *coordinates* of various events, but the *distances* between the points representing those events on the diagram are *not* proportional to the actual spacetime intervals between those events in spacetime.

This is very strange, and it may seem particularly strange that two events (such as A and C in Fig. 4.7b) can occur at different places and times and yet have *zero* spacetime interval between them. Nonetheless, there exists a useful analogy with something you may have seen before. Imagine a map of the world where the lines of longitude and latitude are drawn as equally spaced straight lines (as shown in Fig. 4.8). Have you ever noticed how the shapes and sizes of the continents appear very warped near the north and south poles on such maps? For example, look at the continent of Antarctica on the map shown. It looks huge and seems to be shaped like a strip. But in fact it is not so large, and it has a nearly circular shape: its size and shape are quite distorted by the nature of the map. The shapes of Greenland and northern Canada are distorted as well. As a matter of fact, the two points marked a and b on the map are both at 90° north latitude, i.e., at the north pole. Thus though these points are separated by a significant distance on the map, the physical distance between these points on the surface of the earth is zero!

Why does this map not accurately represent the distances between points on the earth's surface? The problem is that the surface of the earth as a whole is the surface of a *sphere,* which has a very different geometry than the euclidean geometry of a flat sheet of paper. For example, on a sheet of paper, the interior angles of a triangle always add up to 180°, parallel lines never intersect, and so on. But on the surface of the earth, the interior angles of a triangle can add up to *more* than 180° (consider a triangle with one vertex at the north pole and two vertices at the equator), initially paral-

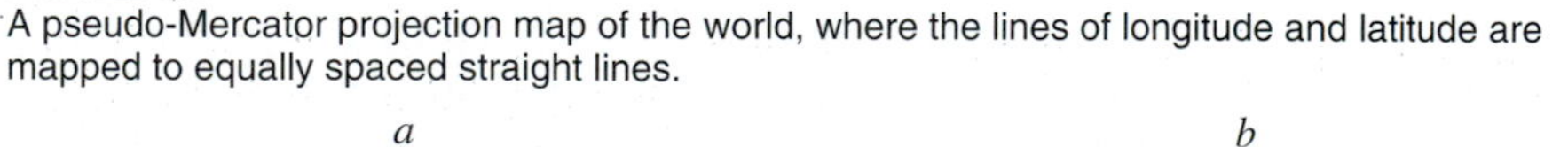

FIGURE 4.8
A pseudo-Mercator projection map of the world, where the lines of longitude and latitude are mapped to equally spaced straight lines.

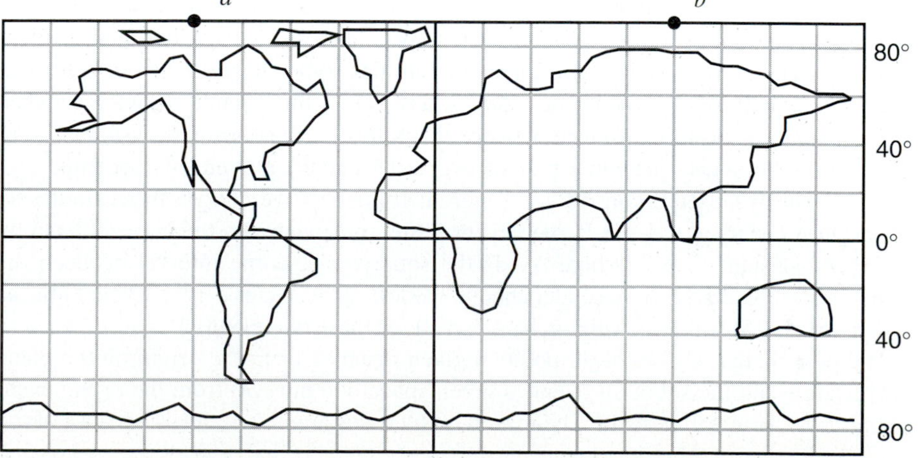

lel lines do not remain parallel, and so on. Because of these fundamental geometric differences between the surface of the earth and the sheet of paper, any flat map of the earth will necessarily be a distorted representation: one *cannot* make a map of the surface of the earth on a flat sheet of paper and have distances on the sheet correspond to actual distances on the earth.

Similarly, one *cannot* draw a spacetime diagram in such a way that distances between points on the drawing are proportional to the spacetime intervals between the corresponding events. Like the surface of the earth, spacetime has a different geometry than the flat sheet of paper on which a spacetime diagram is drawn. The fact that the metric equation of spacetime has negative signs where the corresponding pythagorean relation has positive signs is symptomatic of this difference.

The moral of the story is this: do not expect a spacetime diagram to give you *direct* information about the spacetime interval between various events, any more than you would expect a flat map of the earth to give you accurate information about distances on the earth's surface. A spacetime diagram is meant to visually represent the *coordinates* of events and the worldlines of particles, nothing more. You can always compute the spacetime interval between two events from the coordinates of those events if necessary.

4.6 THE METRIC EQUATION AND THE PYTHAGOREAN THEOREM

In spite of the issue raised in the previous section, we can extend the geometric analogy to spacetime by exploring the basic similarities (as well as differences) in how the metric equation describes the geometry of spacetime and how the pythagorean theorem describes the geometry of a plane.

The most important thing about *both* these equations is that they enable us to calculate an *absolute* quantity (Δs or Δd) in terms of *frame-dependent* coordinate differences measured in an arbitrary inertial frame or coordinate system. This similarity is illustrated in Figs. 4.9 and 4.10. Figure 4.9 shows the *same* pair of points on the plane (A and B) as viewed in coordinate systems having various different orientations with respect to "north." Note that if we set up the coordinate systems so that point A is at the origin, then point B in each coordinate system lies somewhere on the *circle* defined by the equation $x^2 + y^2 = \text{constant} = \Delta d^2$, where Δd^2 is the squared distance between the points (since Δd is the distance between the points in *all* coordinate systems). Similarly, Fig. 4.10 shows the same pair of events (A and B) plotted on spacetime diagrams drawn by observers in different inertial frames. If we choose these frames so that event A occurs at $t = x = 0$, then event B lies somewhere on the curve defined by $t^2 - x^2 = \text{constant} = \Delta s^2$, where Δs^2 is the squared spacetime interval between the events (since Δs has a frame-independent value): such a curve is a **hyperbola,** as shown. (Note that we are assuming $\Delta y = \Delta z = 0$ for these two events.)

The point is that the set of all points a given distance from the origin on the plane form a circle and the set of all events a given spacetime interval from the origin event in spacetime form a hyperbola. The reason both curves are not circles is because of the

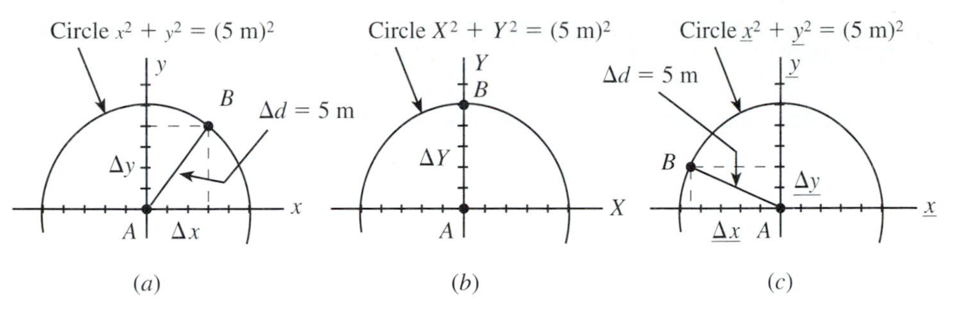

FIGURE 4.9
(*a*) A pair of points on the plane that are separated a distance of 5 m. In this coordinate system, these points happen to have coordinate separations of $\Delta y = 4$ m and $\Delta x = +3$ m. (*b*) We can find a coordinate system in which *A* and *B* lie along the vertical axis ($\Delta X = 0$). In this unique system, the coordinate separation ΔY is equal to the distance between the points. *(c)* If we twist the axes farther clockwise, we can find a coordinate system where the coordinate separations are $\Delta \underline{y} = 2$m and $\Delta \underline{x} = -4.6$ m. In *all* circumstances, though, *B* will lie somewhere on the circle shown.

negative sign in the metric equation that does not appear in the pythagorean relation. But there is a nice one-to-one correspondence between circles in plane geometry and hyperbolas in spacetime geometry.

Note that one consequence of difference between the metric equation and the pythagorean relation, is that in Fig. 4.9 we see that the *y*-coordinate separation between a pair of points is always *less* than or equal to the distance between the points (the hypotenuse on the diagram): $\Delta y \leq \Delta d$. In Fig. 4.10, though, we see that the coordinate time Δt between a pair of events is always greater than the spacetime interval Δs between them: $\Delta t \geq \Delta s$, even though the hypotenuse that *represents* Δs on the diagram *looks* larger.

FIGURE 4.10
(*a*) A pair of events in spacetime that are separated by a spacetime interval of 4 s. In this inertial frame, these events happen to have coordinate differences of $\Delta x = 5$ s and $\Delta x = +3$ s. *(b)* We can find an inertial frame in which *A* and *B* occur at the same place ($\Delta X = 0$). In this unique frame, the coordinate time ΔT is equal to the spacetime interval between the events. *(c)* If we check frames moving even faster to the right, we can find a frame where $\Delta \underline{t} = +6.4$ s and $\Delta \underline{x} = -5$s. In *all* frames, though, *B* will lie somewhere on the hyperbola shown.

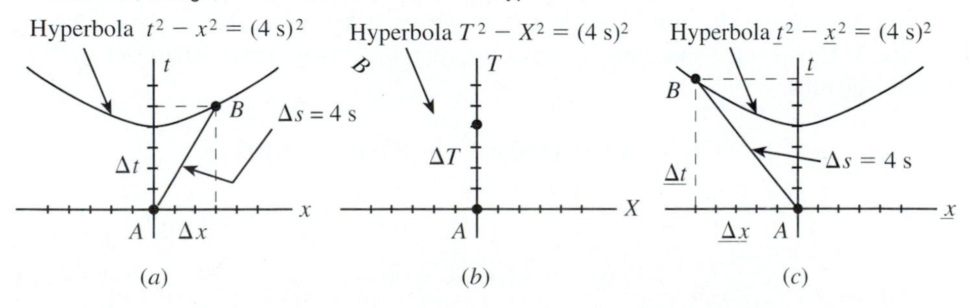

4.7 SOME ELEMENTARY APPLICATIONS OF THE METRIC EQUATION

4.7.1 The Spacetime Interval Is Frame-Independent

Problem A firecracker explodes. A second firecracker explodes 25 ns away and 52 ns later as measured in a certain inertial frame (the Home Frame). In another inertial frame (the Other Frame), the two explosions are measured to occur 42 ns apart in space. How long a time passes between the explosions in the Other Frame?

Solution The key in this problem is to recognize that the spacetime interval between the two explosion events is frame-independent. That is, if we calculate it using the metric equation in the Home Frame, we must get the same answer that we would get if we calculated it in the Other Frame. That is,

$$\Delta t^2 - \Delta d^2 = \Delta s^2 = \Delta t'^2 - \Delta d'^2 \tag{4.6}$$

Solving this equation for the unknown $\Delta t'$, we get

$$\Delta t'^2 = \Delta t^2 - \Delta d^2 + \Delta d'^2 = (52 \text{ ns})^2 - (25 \text{ ns})^2 + (42 \text{ ns})^2 = 3800 \text{ ns}^2$$

Therefore,

$$\Delta t' = \sqrt{3800} \text{ ns} \approx 62 \text{ ns} \tag{4.7}$$

4.7.2 Coordinate Time and the Spacetime Interval

Problem A certain physics professor fleeing the wrath of a set of irate students covers the length of the physics department hallway (a distance of 120 ns) in a time of 150 ns as measured in the frame of the earth. Assuming that the professor moves at a constant velocity, how much time does the professor's watch measure during the trip from one end of the hallway to the other?

Solution Part of the trick in many relativity physics problems is to rephrase a word problem in terms of *events.* In this case, let event A be the professor entering the hallway and event B be the professor's hasty departure from the other end. In the reference frame of the earth, these events occur a time $\Delta t = 150$ ns apart and a distance $\Delta d = 120$ ns apart. The professor's watch, however, is present at each of the events, so the watch registers the spacetime interval between these two events. Therefore, by the metric equation

$$\Delta s^2 = \Delta t^2 - \Delta d^2 = (150 \text{ ns})^2 - (120 \text{ ns})^2 = 8100 \text{ ns}^2$$

Thus

$$\Delta s = \sqrt{8100} \text{ ns} = 90 \text{ ns} \tag{4.8}$$

4.7.3 The Twin Paradox (A First Glance)

Problem A spaceship departs from our solar system and travels at a constant velocity to the star Alpha Centauri 4.3 light-years away, then instantaneously turns around (never mind about the impossible accelerations involved) and returns to the solar system at the same constant speed. Assume that the trip takes 13 years as measured by clocks here on earth. How long does the trip take as measured by clocks on the spaceship?

Solution Again, we need to translate the word problem into a problem about measuring the time between events. Let event A represent the ship's departure from the starbase, event B its arrival at Alpha Centauri, and event C its return to the solar system (see Fig. 4.11). A clock in the spaceship does *not* measure the spacetime interval between events A (departure from the solar system) and C (return to the solar system) even though the clock is present at both events. This is because the clock is accelerated when the spaceship turns around and so is not inertial. To find the total elapsed time registered on the ship clock, we have to consider each leg of the trip separately. The ship's clock *does* measure the spacetime interval between events A and B and also measures the spacetime interval between events B and C, as it is inertial during each individual leg of the trip and is present at the events in question. The total time registered by the ship's clock is thus the sum of the spacetime intervals between A and B and between B and C.

We can use the metric equation to compute these spacetime intervals from the coordinate differences for these events measured in the space station's frame. Events A and B occur $\Delta t = 6.5$ years apart in time and $\Delta d = 4.3$ years apart in space. The spacetime interval between these events is

$$\Delta s_{AB} = \sqrt{\Delta t^2 - \Delta d^2} = \sqrt{(6.5\text{ y})^2 - (4.3\text{ y})^2} \approx 4.9\text{ y} \tag{4.9}$$

The spacetime interval between events B and C is exactly the same. The total elapsed time for the trip as measured by a clock on the ship is thus 2 (4.9 years) = 9.8 years,

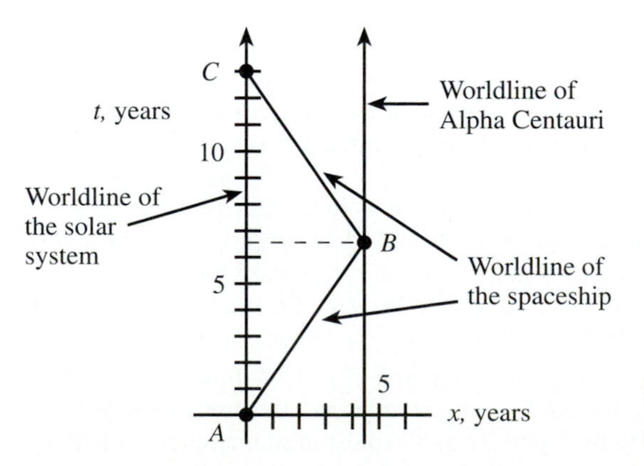

FIGURE 4.11
Worldline of a spaceship traveling to Alpha Centauri and back.

which is somewhat shorter than the time of 13 years measured by clocks in the solar system.

Note that the line on the diagram connecting points A and B *looks* longer than 6.5 years, but the spacetime interval that this line represents is actually *shorter* than 6.5 years. This is a specific illustration of the issue discussed in Sec. 4.5.

PROBLEMS

4.1 THE SPACETIME INTERVAL. In the Home Frame, two events are observed to occur with a spatial separation of 12 ns and a time coordinate separation of 24 ns.

a An inertial clock travels between these events in such a manner as to be present at both events. What time interval does this clock read between the events?

b What is the speed of this clock as measured in the Home Frame?

4.2 THE SPACETIME INTERVAL. An alien spaceship moving at a constant velocity goes from one end of the solar system to the other (a distance of 10.5 h) in 13.2 h as measured by clocks on the earth. What time does a clock on the spaceship read for the passage? (*Hint:* Rephrase in terms of events.)

4.3 THE SPACETIME INTERVAL. A clock travels from one end of the Milky Way galaxy to the other (an approximate distance of 100,000 light-years) at a constant velocity of magnitude $v = 0.999$, as measured in the frame of the galaxy. How much time does the traveling clock register for this trip? (*Hint:* Rephrase this problem in terms of events. How far apart are these events in space and time in the galaxy's frame?)

4.4 A MUON THOUGHT EXPERIMENT. A muon is created by a cosmic-ray interaction at an altitude of 60 km. Imagine that after its creation the muon hurtles downward at a speed of 0.998 as measured by an earth-based observer. After the muon's "internal clock" registers 2.0 μs (which is a bit longer than the average life of a muon), it decays.

a If the muon's internal clock were to measure the same time between its birth and death as clocks on earth do (i.e., special relativity is not true and time is universal and absolute), about how far would this muon travel before it decays?

b How far will this muon *really* travel before it decays?

4.5 THE SPACETIME INTERVAL IS FRAME-INDEPENDENT. In one inertial frame (the Home Frame), two events are observed to occur at the *same place*, but $\Delta t = 32$ ns apart in time. In another inertial frame (the Other Frame), the same two events are observed to occur 45 ns apart in space.

a What is the coordinate time interval between the events in the Other Frame?

b Compute the speed of the Home Frame as measured by observers in the Other Frame. (*Hint:* The events occur at the *same place* in the Home Frame. So how far does the Home Frame appear to move in the time between the events as seen in the Other Frame? What is the time between the events in the Other Frame?)

4.6 AN EXPLODING ROCKET. The spacetime diagram in Fig. 4.12 shows the worldline of a rocket as it leaves the earth, travels for a fixed time, and then explodes. What is the elapsed time between the rocket's departure and explosion as measured by a clock on the rocket? What is the spacetime interval between these two events?

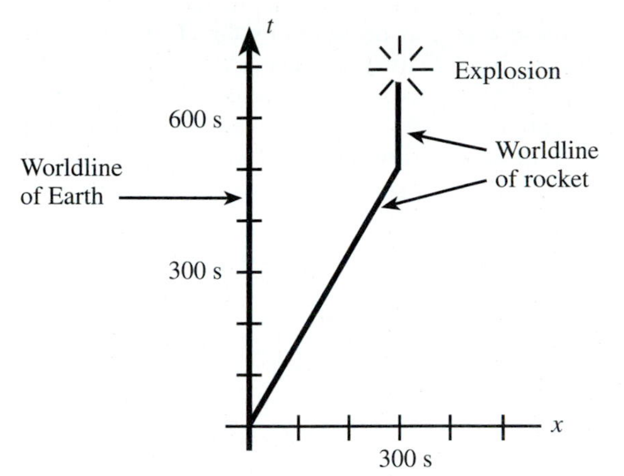

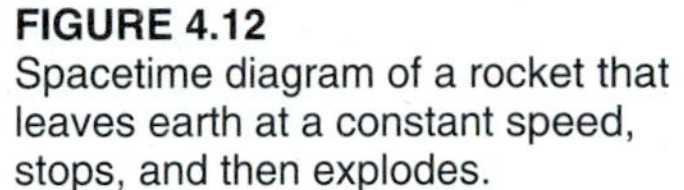

FIGURE 4.12
Spacetime diagram of a rocket that leaves earth at a constant speed, stops, and then explodes.

4.7 RAPID TRANSIT. A new rapid transit line advertises they will carry you from Los Angeles to Seattle (a distance of about 3000 km) at such a high rate of speed that your watch only registers half the time for the trip that synchronized clocks in the station read.

a What is the approximate distance between Los Angeles and Seattle in microseconds?

b What time interval must the synchronized station clocks register between departure from Los Angeles and arrival in Seattle if the advertisement is to be true?

c What is the approximate speed of the train?

4.8 PARTICLE DECAY. Imagine that a certain unstable subatomic particle decays with a half-life at rest of about 2.0 μs (that is, if at a certain time you have N such particles at rest, about 2.0 μs later you will have $N/2$ remaining). Now imagine that with the help of a particle accelerator, we manage to produce in the laboratory a beam of these particles traveling at a speed $v = 0.866$ in SR units (as measured in the laboratory frame). This beam passes through a detector A, which counts the number of particles passing through it each second. The beam then travels a distance of about 2.08 km to detector B, which also counts the number of particles passing through it (see Fig. 4.13).

a Let Event A be the passing of a given particle through detector A and event B be the passing of the same particle through detector B. How much time will a laboratory observer measure between these events? (*Hint:* No relativity needed!)

FIGURE 4.13
Laboratory setup for a particle decay experiment.

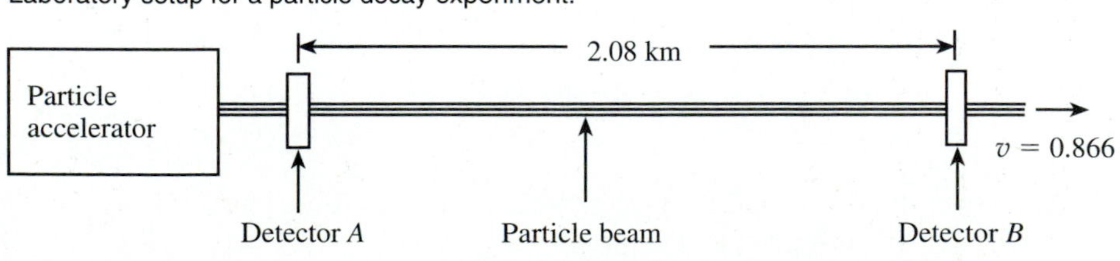

b How much time passes between these events as measured by the clock inside the particle, according to relativity theory? If relativity is true, about what fraction of the particles that pass through detector *A* survive to pass through detector *B*?

c According to the newtonian conception of time, the time measured by a particle clock between the events would be the *same* as the time measured by lab clocks. If this were so, what fraction of the particles passing through detector *A* survive to detector *B*?

4.9 SPACETIME TRIGONOMETRY. The *hyperbolic sine* and *hyperbolic cosine* functions of a quantity θ are defined as follows:

$$\sinh\theta = \tfrac{1}{2}(e^{\theta} - e^{-\theta}), \qquad \cosh\theta = \tfrac{1}{2}(e^{\theta} + e^{-\theta}) \tag{4.10}$$

a Prove that $\cosh^2\theta - \sinh^2\theta = 1$. This means that if the spacetime interval between two events occurring along the spatial x axis is Δs, then the coordinate separations Δt and Δx between these events can be written $\Delta t = \Delta s\cosh\theta$ and $\Delta x = \Delta s\sinh\theta$ for some appropriately chosen value of θ (see Fig. 4.14).

b Argue that as θ goes to zero, $\sinh\theta \to 0$ while $\cosh\theta \to 1$, just like the corresponding trigonometric functions.

c Argue that θ in the hyperbolic case is *not* the angle that the line *AB* makes with the *t* axis in the spacetime diagram of Fig. 4.13*b*. Argue in fact that as $\theta \to \infty$, the angle that *AB* makes with the *t* axis approaches 45°.

d Argue that if v is the speed of an object that goes from event *A* to event *B* at a constant velocity, the angle θ is in fact $\tanh^{-1} v$.

e When $v = 0.80, \theta = 1.10$ (if your calculator can do inverse hyperbolic functions, verify this). What are the values of $\cosh\theta$ and $\sinh\theta$ for this value of θ? (Use the definitions of these functions given above if your calculator cannot evaluate hyperbolic functions.) When $v = 0.99, \theta = 2.65$ (again, verify if you can). What are the values of $\cosh\theta$ and $\sinh\theta$ for this value of θ?

FIGURE 4.14
(a) Plane trigonometry. *(b)* Spacetime trigonometry. (Note that $\Delta s\cosh\theta > \Delta s$, even though the "hypotenuse" representing Δs *looks* bigger on the diagram.)

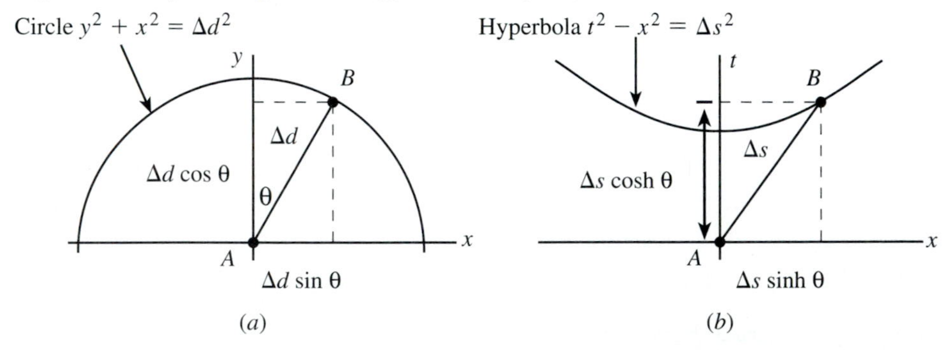

5

PROPER TIME

The great discoverers can readily be classed under two types of mentality: those who dig deep and those who range wide. Those who possess the gift of combining depth with breadth are rare indeed. Einstein was one of them.

François le Lionnais[1]

5.1 INTRODUCTION

In the last chapter, we derived the metric equation, which links the spacetime coordinate differences between two events measured in *any* inertial frame with the frame-independent spacetime interval between those events. In this chapter we will use the metric equation to calculate the proper time that a clock traveling along various worldlines might measure between events along its worldline.

In a given inertial reference frame, the motion of a clock (and thus its worldline through spacetime) can be completely specified by stating its initial position and then describing its velocity $\vec{v}(t)$ as a function of time subsequently. What we want to do in this chapter is express the proper time that would be measured by that clock between events along its worldline in terms of the clock's velocity $\vec{v}(t)$ in a given inertial frame and the coordinate time Δt between those events in the same frame. We begin in Sec. 5.2 with the simplest case where $\vec{v}$ is constant and then move on in Sec. 5.4 to deriving a more general formula valid for any worldline.

[1]Quoted in A. P. French (ed.), *Einstein: A Centenary Volume,* Cambridge, Mass.: Harvard, 1979, p. 133.

The remainder of the chapter is devoted to the application of these formulas to various kinds of situations, including a study of real experimental results, the Doppler shift, and the famous "twin paradox."

5.2 PROPER TIME ALONG A STRAIGHT WORLDLINE

The metric equation links the spacetime interval Δs between two events (call them A and B) with the coordinate differences Δt, Δx, Δy, Δz measured between these events in any given inertial reference frame (call this the Home Frame):

$$\Delta s^2 = \Delta t^2 - \Delta x^2 - \Delta y^2 - \Delta z^2 \tag{5.1}$$

We can express this equation in a different form if we divide both sides by Δt^2:

$$\frac{\Delta s^2}{\Delta t^2} = 1 - \frac{\Delta x^2}{\Delta t^2} - \frac{\Delta y^2}{\Delta t^2} - \frac{\Delta z^2}{\Delta t^2} = 1 - \frac{\Delta d^2}{\Delta t^2} \tag{5.2}$$

Now, Δd is the spatial distance between the two events as measured in the Home Frame, and Δt is their separation in time. The clock that measures Δs must be present at both these events by definition, so in the Home Frame this clock will be observed to cover the distance Δd between the events in the time Δt. This means that speed (i.e., the magnitude of this clock's constant velocity vector), as determined by observers in the Home Frame, must be $v = \Delta d/\Delta t$. Plugging this into Eq. (5.2) and taking the square root of both sides and multiplying by Δt, we get

$$\Delta s = \sqrt{1 - v^2}\,\Delta t \tag{5.3}$$

This very useful equation links the spacetime interval Δs measured by an inertial clock present at two events with the coordinate time separation Δt between those events in some inertial frame and the speed v of the clock *as measured in the same inertial frame*. One can see several things immediately from this equation. (1) The spacetime interval between two events is always *less* than (or at best equal to) the coordinate time measured between those events in any given reference frame. (2) As the speed of the clock measuring the spacetime interval approaches 1 (the speed of light) in the inertial frame, the discrepancy between the spacetime interval and the coordinate time between the events can become quite large. (3) Conversely, if the speed of the clock present at both events is *small* compared to the speed of light ($v << 1$), that clock will register almost the same time between them as measured in the inertial frame: $\Delta s \approx \Delta t$.

When applying this equation, it is important to remember two things. First of all, coordinate time Δt and spacetime interval Δs represent the time interval between two events measured in two *fundamentally different* ways (just as the northward displacement Δy and the distance Δd represent two fundamentally different ways of measuring the spatial separation of two points on the earth's surface). The coordinate time between events is measured with a pair of synchronized clocks in an inertial frame, while the spacetime interval is measured by an inertial clock present at both events. One cannot use the equation to link readings on just any old set of clocks.

Second, the quantities Δs and Δt appearing on both sides of $\Delta s = \sqrt{1 - v^2}\,\Delta t$ in fact refer to the time interval between *a specific pair of events* measured in two different ways. Perhaps the most common error made by beginners in applying Eq. (5.3) is implicitly using *different* pairs of events to delimit the time intervals Δs and Δt. To avoid making this error, you must always think carefully about *exactly what events* delimit Δs and Δt.

Note that in this case the spacetime interval Δs between the events is the *same* as the proper time $\Delta\tau$ that the inertial clock present at both events registers between the events (by definition). Equation (5.3) thus allows us to compute the proper time between the events in the special case where the clock moves between the events at a constant velocity.

5.3 A USEFUL APPROXIMATION FOR SMALL SPEEDS

The square root that appears in Eq. (5.3) is rather difficult to evaluate for very small speeds ($v << 1$). The speeds of objects that we encounter on an everyday basis are on the order of $v = 10^{-8}$ in SR units, meaning $v^2 \approx 10^{-16}$. When you try to evaluate the square root $\sqrt{1 - v^2}$ in such a case, your calculator usually simply returns 1.0, since few calculators keep track of enough decimal places to accurately register the subtraction of 10^{-16} from 1. But in such cases, we can make an approximation that helps us convert the square root to a more usable form.

It can be shown (see Prob. 5.12) that

$$(1+x)^a \approx 1 + ax \qquad \text{if } |x| << 1 \tag{5.4}$$

This equation, called the **binomial approximation,** is a *very* useful approximation with a wide range of applications in physics (memorize it!). In the case that we are considering here, if we identify $x \equiv -v^2$ and $a = \frac{1}{2}$, we find that

$$\sqrt{1 - v^2} \approx 1 + \tfrac{1}{2}(-v^2) = 1 - \tfrac{1}{2}v^2 \qquad \text{if } v^2 << 1 \tag{5.5}$$

You can check the validity of this approximation with a calculator (try $v = 0.1, 0.01$, etc.). An example of the use of this formula follows.

Problem Imagine that you are driving in a straight line along the freeway at a constant speed of 30 m/s. As you pass a certain bank sign, you notice that it changes from reading 2:59 P.M. to exactly 3:00 P.M. (event A). You check your dashboard clock at that time and see that it changes at the same instant. Some kilometers down the road, you see another bank sign change from reading 3:09 P.M. to 3:10 P.M. (event B). Assuming that the two bank signs are synchronized in the inertial frame of the ground, what is the elapsed time between A and B according to the dashboard clock?

Solution Your dashboard clock is present at both events A and B and so registers a proper time between these events. Since you are moving in a straight line at a constant velocity, this proper time is in fact the spacetime interval Δs between the events. The pair of synchronized bank clocks register the coordinate time ($\Delta t = 10$ min $= 600$ s) between the events in the ground frame. In SR units, your speed in the ground frame is $(30 \text{ m/s})/(3.0 \times 10^8 \text{ m/s}) = 1.0 \times 10^{-7}$ which is << 1. So applying Eq. (5.3) and the binomial approximation, we see that the time measured by your dash clock is

$$\begin{aligned}\Delta s &= \sqrt{1 - v^2}\,\Delta t \approx (1 - \tfrac{1}{2}v^2)\,\Delta t = \Delta t - \tfrac{1}{2}v^2\,\Delta t \\ &= 600 \text{ s} - 0.5(1.0 \times 10^{-7})^2(600 \text{ s}) = 600 \text{ s} - 3.0 \times 10^{-12} \text{ s} \\ &= 599.999999999997 \text{ s}\end{aligned} \tag{5.6}$$

(Since most calculators do not keep track of 14 digits, you have to use the binomial approximation and then do the subtraction by hand, as I have done.) Note that this is a totally insignificant difference: when the relative speed of a clock and a reference frame is as small as is typical in daily life, the difference between the clock's proper time between two events and the coordinate time between the same events is completely negligible. This is why the concept of "universal and absolute time" seems so natural!

5.4 A CURVED FOOTPATH

Equation (5.3) compares the coordinate time Δt between two events as measured in any given inertial reference frame with the proper time $\Delta\tau$ measured by an *inertial* clock present at both events ($\Delta\tau = \Delta s$ in this case). In this section and the next, we will generalize Eq. (5.3) to connect coordinate time to the proper time $\Delta\tau$ measured by any clock present at both events, inertial or not.

It has sometimes been said that one must use the theory of general relativity to properly analyze the behavior of accelerating clocks. In fact, we can quite adequately analyze the behavior of such clocks using only the metric equation if we simply remember the analogy between the proper time between events in spacetime and the pathlength between two points on a plane.

Consider a footpath around a small pond, which is illustrated in Fig. 5.1*a* by a scale drawing with a superimposed coordinate system. We could measure the length of this path from point A to point B with a long, flexible tape measure. But once we have set up a coordinate system, we can also *compute* the length of the path in the following manner. Imagine dividing up the path into a large number of infinitesimally small sections, as shown in Fig. 5.1*b*.[1] If we make these sections small enough, each will be approximately straight.

In this limit, the pathlength dL of a given segment as measured by a flexible tape

[1] The analogy with pathlength presented here follows E. F. Taylor and J. A. Wheeler, *Spacetime Physics*, San Francisco: Freeman, (1966), pp. 32–34.

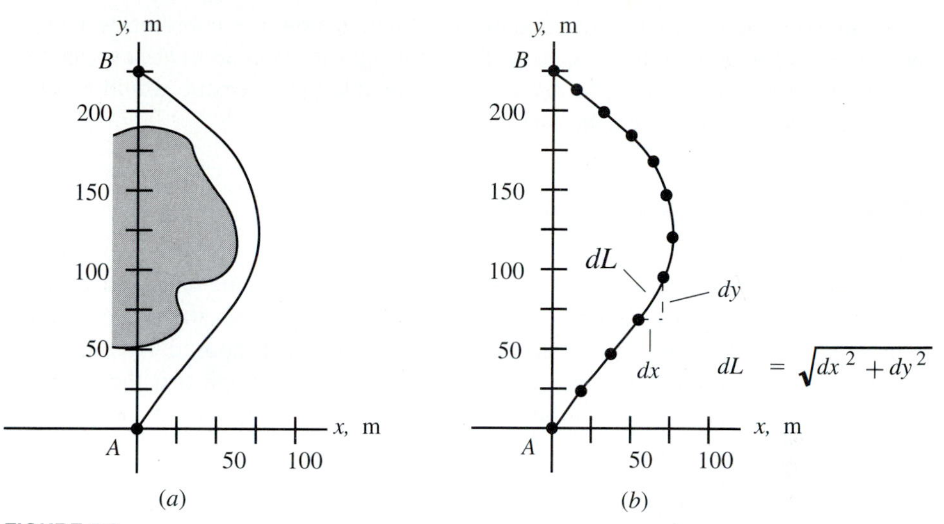

FIGURE 5.1
(*a*) Map of a path around a small pond, with a superimposed coordinate system shown. (*b*) We can compute the length of the path by subdividing the path into many infinitesimal, almost straight sections, finding the length *dL* of each section, and summing to find the total length.

measure will be almost equal to the straight length computed (using the pythagorean theorem) from the coordinate differences of the segment's endpoints:

$$dL^2 \approx dx^2 + dy^2 \qquad \text{or} \qquad dL = \sqrt{dx^2 + dy^2} \tag{5.7}$$

The total length L_{AB} of the path from A to B is the sum of all the segment lengths, which in the infinitesimal-segment limit becomes the integral

$$\Delta L_{AB} = \int_{\text{path}} dL = \int_{\text{path}} \sqrt{dx^2 + dy^2} \tag{5.8}$$

Note that since the length dL of each segment is greater than its northward extension dy, the total pathlength between points A and B will be greater than the straight-line northward distance of 250 m between A and B; quite generally, therefore, $\Delta L_{AB} \geq \Delta d_{AB}$.

If we think of the path as being specified by giving the x coordinate of each point on the path as a function of y [that is, the path is specified by the function $x(y)$], then we can write the integral above as a single-variable integral over x by pulling a factor of dy out of the square root:

$$L_{AB} = \int_{y_A}^{y_B} \sqrt{1 + \left(\frac{dx}{dy}\right)^2}\, dy \tag{5.9}$$

[Note that we are considering y to be the independent variable and x to be the dependent variable in the equation above. This reversal of convention is necessary because $y(x)$ is not well-defined for a function of the shape shown in the diagram.]

As we have discussed before, though this equation uses the coordinates x and y measured in a given coordinate system, the pathlength itself is an invariant quantity: we will get the same answer (the answer that a flexible tape measure would give) no matter what coordinate system we use.

5.5 CURVED WORLDLINES IN SPACETIME

The analogy to worldlines in spacetime is direct. Consider the worldline of a particle that travels out from the origin of some inertial reference frame a certain distance along the x axis and then returns. Such a worldline is shown on the spacetime diagram in Fig. 5.2*a* with the coordinate axes of that frame superimposed. Such a worldline describes an accelerating particle; we can see from the graph that the particle's x velocity $v_x = dx/dt$ (which is the inverse slope of its worldline on the diagram) changes as time progresses.

A clock traveling with the particle measures the proper time $\Delta\tau_{AB}$ between events A and B along this worldline (by definition of proper time). But once we have measured the worldline of the particle in an inertial reference frame (*any* inertial frame), we can *calculate* what this clock will read between events A and B by using the metric equation in a manner analogous to our determination of the pathlength between points A and B in the previous section.

Imagine that we divide the particle's worldline up into many infinitesimal segments, each of which is nearly a straight line on the spacetime diagram (Fig. 5.2*b*). That is, each segment is chosen to be short enough so that the particle's *velocity* is

FIGURE 5.2
(*a*) Spacetime diagram of the motion of a particle's worldline based on measurements obtained in some inertial frame. (*b*) Computing the proper time along the worldline by subdividing it into many infinitesimal and almost straight segments, finding the proper time $d\tau$ elapsed during each segment, and then summing to find the total proper time.

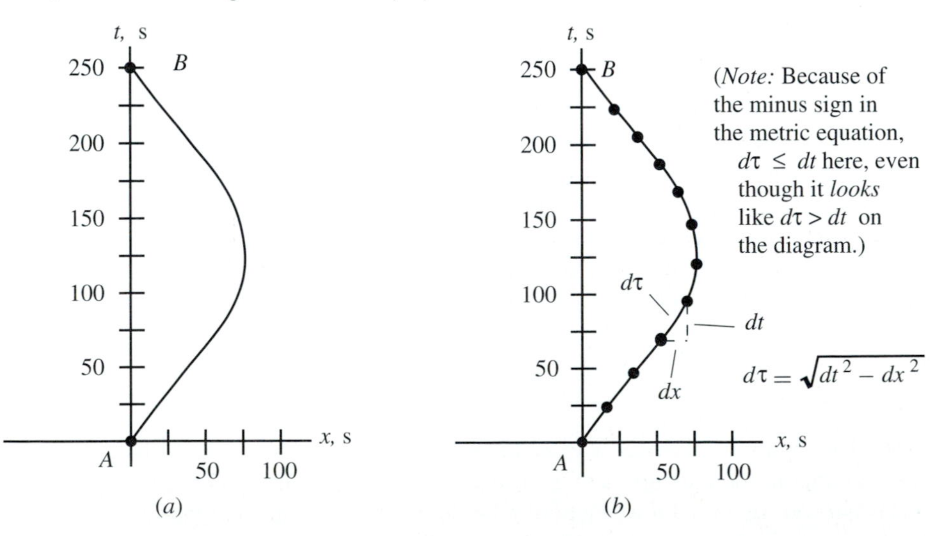

approximately constant as it traverses that segment. If a given segment is short enough so that the particle's velocity is almost constant along it, the proper time $d\tau$ that a clock would measure along that segment will be almost equal to the spacetime interval ds between the events that mark the ends of the segment, since the clock is present at both these events and travels between them with an *almost* constant velocity. Therefore, by the metric equation

$$d\tau^2 \approx ds^2 = dt^2 - dx^2 - dy^2 - dz^2 \tag{5.10}$$

Taking the square root and pulling out a factor of the coordinate time dt, we get

$$d\tau = \sqrt{dt^2 - dx^2 - dy^2 - dz^2} = \sqrt{1 - \left(\frac{dx}{dt}\right)^2 - \left(\frac{dy}{dt}\right)^2 - \left(\frac{dz}{dt}\right)^2}\, dt = \sqrt{1 - v^2}\, dt \tag{5.11}$$

In some sense, Eq. (5.11) is just a differential version of Eq. (5.3), but its meaning is a bit more general. It expresses the infinitesimal proper time $d\tau$ measured by a clock traveling between two *infinitesimally separated* events in terms of the coordinate time dt between those events measured in some inertial frame and the clock's instantaneous speed v measured in that frame. The clock may be moving along *any* smooth worldline (with v not necessarily constant).

To find the total proper time measured between events A and B by a clock traveling along the worldline, we sum up the proper times measured for each segment of the worldline, which in the infinitesimal-segment limit amounts to integrating Eq. (5.11):

$$\Delta\tau_{AB} = \int_{\text{worldline}} \sqrt{1 - v^2}\, dt \tag{5.12}$$

If the speed v of the clock is expressed as a function of time $v(t)$, then the integral is simply an ordinary one-variable integral with respect to t, which can be evaluated in principle.

This equation links the total proper time $\Delta\tau_{AB}$ between two events measured by a clock traveling between those events with t and v, the coordinate time between the events and the speed of the clock, both as measured in some given (but arbitrary) inertial reference frame. Though we are using an inertial frame to measure t and v, remember that the result of Eq. (5.12) is frame-independent, since the proper time is measured by the clock in question *directly* without reference to any frame.

Note that since $\sqrt{1 - v^2}$ is always ≤ 1, the proper time measured between two events along any path will be smaller than or equal to the coordinate time between those events measured in any inertial frame: $\Delta\tau_{AB} \leq \Delta t_{AB}$ under all circumstances.

If the speed v of the clock is *constant*, the integral in Eq. (5.12) can be done very easily:

$$\Delta\tau_{AB} = \sqrt{1 - v^2} \int_{t_A}^{t_B} dt = \sqrt{1 - v^2}\, \Delta t_{AB} \qquad \text{if } v = \text{constant} \tag{5.13}$$

Please note that "constant speed" here does not necessarily imply "constant velocity," as the *direction* of a particle's velocity may change without changing its speed. The equation can be applied to clocks traveling along straight or curved worldlines, as long as the speed of the clock remains fixed. Equation (5.12) may be used when the speed changes.

Note that Eqs. (5.12) and (5.13) [and Eq. (5.3) as well] break down if $v > 1$: in such a case they predict that the time registered by the traveling clock is an *imaginary* number (which is even worse than being negative!). Remember that these equations are all based on the metric equation, whose derivation (see Sec. 4.2) is only valid for pairs of events for which $\Delta t > \Delta d$, that is, events between which it is possible to send a clock traveling with $v < 1$. Therefore, the equations presented so far do not specify what a clock traveling faster than the speed of light would read between two events. We will see later that the principle of relativity in fact implies that it is *impossible* for a clock to travel faster than the speed of light in any reference frame. The failure of these equations for the case where $v > 1$ is our first indication of this basic truth.

5.6 THE SPACETIME INTERVAL IS THE LONGEST PROPER TIME

Note that Eq. (5.12) implies that generally the proper time measured by a clock between two events will indeed depend on the worldline that the clock follows between the events: specifically, the proper time depends on the particular way that the speed v of the clock varies with time. This is analogous to the way that the pathlength between two points on a plane depends on the curvature of the path along which it is measured.

In euclidean geometry, the straight-line distance between two points is the *shortest* possible pathlength between the two points. In this section, we will prove that an inertial clock that travels between two events (which thus measures the spacetime interval between them) measures the *longest* proper time between those events, i.e., a longer time than any noninertial clock.

Theorem

The worldline of an inertial particle represents the path of *greatest* proper time between two events in spacetime: $\Delta s \geq \Delta\tau$ for all possible worldlines between the events.

Proof Consider an arbitrary pair of events, A and B. The time between these events is measured by two clocks that are present at both events: Clock I follows an *inertial* worldline between the events, while clock NI follows a noninertial worldline. Since clock I is inertial, it can be used to define a Home Frame in which it is at rest. Since it is inertial, its proper time $\Delta\tau_I$ will be equal (by definition) to the spacetime interval Δs_{AB} between the events. We will take advantage of the fact that we can calculate the proper time for a given worldline using *any* inertial reference frame we please: the result will be frame-independent. Let us choose to evaluate the proper times for clocks I and NI in the particular frame where clock I is at rest.

To compute the proper time along any path from event A to event B in this frame, we integrate the factor $\sqrt{1-v^2}$ (with v measured in the Home Frame) from t_A to t_B (also as measured in the Home Frame). For the inertial clock I, the integrand

$\sqrt{1 - v^2} = 1$, since that clock is at rest in the reference frame we are using (so $v = 0$). Since clock *NI* travels along a *different* worldline by hypothesis, it must at least *sometimes* have a nonzero velocity in our chosen frame, implying that $\sqrt{1 - v^2} < 1$ for at least part of the range of the integration. Therefore, $\Delta s_{AB} = \Delta\tau_I = \int [1]\, dt > \int \sqrt{1 - v^2}\, dt = \Delta\tau_{NI}$ (since both integrals have the same endpoints). Since the values of the proper times are frame-independent, the same inequality must apply no matter what inertial frames we use to make the calculation.

Note that in a spacetime diagram based on measurements made in an inertial reference frame, an inertial clock will have a straight worldline (since it moves with constant velocity with respect to any inertial frame) whereas a noninertial clock will have a curved worldline. This theorem thus says that a *straight* worldline between any two events on a spacetime diagram is the worldline of greatest proper time between the events.

That the spacetime interval represents the *longest* proper time between events while the distance represents the *shortest* pathlength between points on a plane is a direct consequence of the negative signs that appear in the metric equation while only positive signs appear in the corresponding pythagorean relation. This is another of the basic differences between spacetime geometry and the euclidean geometry of points on a plane. Even so, there remains a similarity in that a straight worldline (or path) leads to an *extreme* value for the proper time (or pathlength).

Equation (5.3) also implies that the coordinate time Δt between two events measured in *any* inertial reference frame is greater than (or equal to) the spacetime interval Δs between the points. The three kinds of time interval that you can measure between two events thus stand in the strict relation

$$\Delta t \geq \Delta s \geq \Delta\tau \tag{5.14}$$

where the first inequality becomes an equality if the events occur at the *same place* in the inertial reference frame where Δt is measured (so that a clock in that frame can be present at both events) and the second inequality becomes an equality if the clock measuring the proper time follows an *inertial path* between the events (so that its proper time *is* the spacetime interval as well).

5.7 EXPERIMENTAL EVIDENCE

Equation (5.12) implies that if two clocks are synchronized at event A and then travel to event B along different worldlines, they will in general *not* be synchronized when they arrive at event B, since the speeds of the clocks (as measured in some inertial frame) will not generally be the same as they follow their different worldlines. This prediction severely conflicts with our newtonian intuition about time but is a direct (and testable) consequence of the principle of relativity.

In one experiment,[1] muons were generated using a particle accelerator and then kept in a circular storage ring that they traveled around at a measured constant speed

[1] J. Bailey et al., *Nature,* **268,** 1977.

$v = 0.99942$. Though the muons traveled with a constant speed, their *velocity* was not constant because they were traveling in uniform circular motion. In fact, the worldline of such a muon is quite curved and looks something like that shown in the spacetime diagram of Fig. 5.3.

Consider a clock at rest in the laboratory at one point on the storage ring. A pulse of muons passes that clock (call this event A). After going once around the ring, the pulse passes that clock again (call this event B). Both the lab clock and the muons measure proper times between events A and B, since both are present at both events A and B. But the clock and the muons travel along very different worldlines to get there. The clock travels along a straight worldline of constant *velocity* ($\vec{v} = 0$ in the lab frame) and thus measures the spacetime interval between the events. The muons travel along a curved worldline of constant *speed* ($v = 0.99942$ in the lab frame) but with ever-changing velocity. Equation (5.13) can be used in *both* cases to compute the time measured.

The lab clock case is easy: $v = 0$ for this clock in the lab frame so that the proper time that it reads between events A and B is simply

$$\Delta\tau_{AB}\,(\text{lab}) = \sqrt{1-0^2}\,\Delta t_{AB} = \Delta t_{AB} \tag{5.15}$$

where Δt_{AB} is the coordinate time between the events in the lab frame (note that this is also the spacetime interval between the events). A clock traveling with a muon, however, reads

$$\Delta\tau_{AB}\,(\text{muon}) = \sqrt{1-(0.99942)^2}\,\Delta t_{AB} \approx (1/29.3)\Delta t_{AB} \tag{5.16}$$

meaning that a clock traveling with the muon should measure a little more than 1/30 the time between A and B that the laboratory clock does!

As we have discussed before, the fact that muons decay with a specific half-life makes them effectively little clocks, and measuring the rate at which they decay amounts to reading those clocks. If their clocks measure 1/30 of the time between events A and B that an equivalent clock would at rest in the laboratory, then the muons

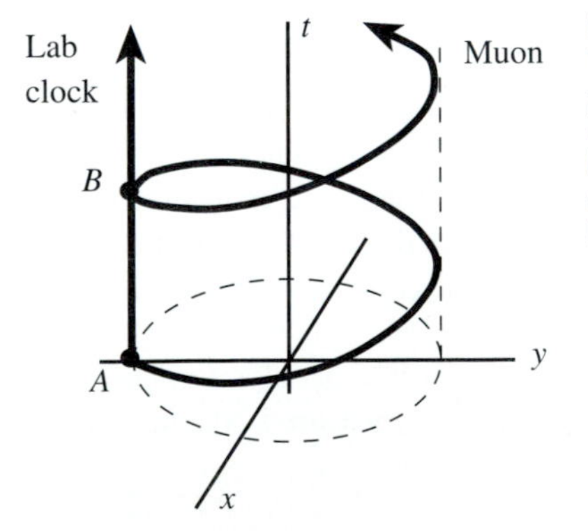

FIGURE 5.3
The curved worldline of a muon traveling in a circular path in a storage ring, as plotted in the laboratory frame of reference. The vertical worldline drawn is the worldline of a clock at rest in the laboratory. Both the lab clock and muon measure (different) proper times between events *A* and *B*.

in the storage ring will be seen to decay roughly 30 times slower than they would at rest in the laboratory. Since the half-life of a muon at rest is known to be 1.52 μs, all that remains to be done is to measure the half-life of the muons in the storage ring. This was done. The muons in the storage ring were measured to have a half-life that was 29.3 times longer than their rest half-life (within experimental uncertainties), in complete agreement with Eq. (5.16).

As a point of interest, the acceleration of these muons as measured in the lab frame is about 10^{15} times the acceleration of gravity! Therefore, Eq. (5.13) (at least) is seen to apply even in cases of extreme acceleration.

In a less extreme experiment performed by J. C. Hafele and R. E. Keating in 1971,[1] a pair of very accurate atomic clocks were synchronized, and then one was put on a jet plane and sent around the world. Upon its return, the jet clock (which followed a non-inertial worldline in its trip around the earth) was compared with the inertial clock that remained at rest in the frame of the earth. With suitable corrections for the effects of gravity (predicted by the general theory of relativity), the results were found to be in complete agreement with Eq. (5.12).

Many other experiments have been performed to check this prediction of the theory of relativity, and all have been in complete agreement with the predictions of Eq. (5.12). Outrageous as it may seem, the idea that two clocks present at the same two events need *not* register the same time between those events is a well-established experimental fact.

5.8 THE TWIN PARADOX

As a result of misinterpreting the meaning of Eq. (5.3) [or (5.12)], many people (including competent professors of physics) have been unnecessarily perplexed by apparent *paradoxes* of relativity theory. One of the most famous of these is called the *twin paradox* (or sometimes the *clock paradox*). This problem generated reams of journal articles (as late as into the 1960s) before the inadequacy of the language and concepts commonly used to describe relativity at that time was sufficiently well-understood.

Here is a statement of the apparent paradox. Andrea and Bernard are twins. When they are both 25 years old, Andrea accepts a commission to be an exobiologist on a expedition to Tau Ceti, a star nearly 8 light-years away from earth. So she flies away on a ship that is capable of near-light speeds, leaving her brother Bernard on earth. The years roll by and the world waits. Finally, hurtling out of the emptiness of space, the spacecraft returns.

As he waits for his sister to emerge, Bernard (now a distinguished-looking man of 50) muses on the bit of relativity that he remembers from college. "$\Delta\tau_{AB} = \int \sqrt{1-v^2}\,dt$," he recalls. Since Andrea has been moving with a large speed v for much of the trip, Andrea's clocks should measure much less time for the trip than his clocks register. This includes biological as well as mechanical clocks, and so Bernard expects to see a

[1] *Science,* **117,** 168, July 14, 1972.

substantially younger sister emerge from the hatch, still displaying their once common youthful vitality. Bernard chews his lip, wondering what it will be like to have a younger "twin" sister.

Similar thoughts run through Andrea's mind as she prepares to disembark. In Andrea's frame of reference, however, she and the spacecraft were motionless, and it is the earth (and thus Bernard) that has moved backward 8 light-years and returned. Andrea (who had the same course in college) thinks that since it is Bernard whose speed $v \neq 0$, it will be *Bernard* who will be younger.

The paradox is clear: each expects the other to be younger from their partial recollection of the relativity theory. To this confusion, we can add a third perspective. The principle of relativity states that the laws of physics are the same in every inertial frame. This means that there is no way of making a physical distinction between two inertial frames: if you perform identical experiments within each reference frame, you must get the same results. But is not the aging process essentially a physical experiment that each person performs in his or her reference frame? If *either* twin is younger than the other, will that not distinguish between the frame of the earth and the frame of the spaceship, contrary to the principle? So perhaps they should have the same age?

We have in fact *already* resolved this paradox in this chapter: we simply need to rephrase it in more appropriate language. The first task is to clarify what *events* we are talking about. Let us define event A to be the departure of the ship from earth. Let its arrival back on earth be event B. Both twins are present at both events A and B, so their clocks (including their biological clocks) measure proper time between these events. The question is, which of these twins measures the longer proper time between these events? To find this out, we need to sketch their worldlines as measured in some inertial reference frame.

For our master inertial reference frame, let us choose a frame at rest with respect to the sun. Since the sun is freely falling around a very distant galactic center of mass, this will be an excellent approximation to an inertial frame. Let us sketch the twins' worldlines as measured in this frame.

Andrea takes off from the earth at event A, which we can take to be the origin event (i.e., the event that defines $t = x = 0$). Her spacecraft travels slowly at first but gradually picks up speed as the spacecraft strains toward the speed of light. Finally, after a few years, the spacecraft reaches cruising speed. But long before it reaches Tau Ceti, it must begin to slow down, lest it flash by the star at some outrageous velocity. Finally, it coasts into the star system and lands. The process of acceleration and deceleration repeats on the way home. Event B is the event of her return to earth. The resulting worldline is drawn on the spacetime diagram of Fig. 5.4.

On this diagram, the earth's (and thus Bernard's) worldline is a tiny helix winding around the t axis, twisting around it about 25 times in the roughly 25-year duration of the flight. So Bernard does *not* measure coordinate time between events A and B, as his vague argument seems to suggest; rather he measures a proper time between those events that is somewhat different than coordinate time. But since Bernard is never more than about 8 light-minutes ($\approx 1.5 \times 10^{-5}$ light-years) from the sun, the squiggles of his little helix are far too tiny to show up on the diagram: his worldline is essentially a straight line up the t axis. Moreover, since Bernard's speed in his worldline (roughly

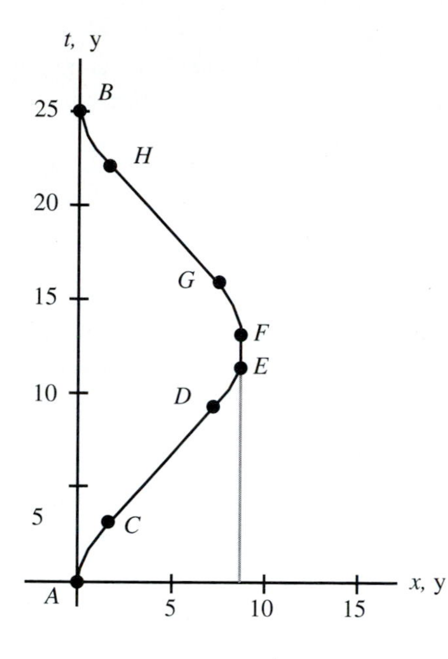

FIGURE 5.4
Andrea's worldline in the reference frame of the sun. *A:* departure from Earth; *C:* spaceship reaches cruising speed; *D:* spaceship begins deceleration; *E:* arrival at Tau Ceti; *F:* departure from Tau Ceti; *G:* spaceship reaches cruising speed; *H:* spaceship begins deceleration; *B:* spaceship arrives at Earth.

equal to the orbital velocity of the earth) is about 30 km/s $\approx 10^{-4}$ in SR units, the proper time measured by his clocks only differs from that measured by coordinate clocks in the sun's frame by about $(1 - \sqrt{1 - v^2})(25\ y) \approx (5\times10^{-9})(25\ y) \approx 4$ s over the time period between events A and B. We see for Bernard that the proper time he measures between events A and B is *essentially* the same as the time measured in the sun's frame between those events.

But for Andrea, the situation is different. In the sun's frame, Andrea spends quite a bit of time traveling at nearly the speed of light: her average speed in this frame is (16.0 ly/25 y) ≈ 0.64. Therefore the factor $\sqrt{1 - v^2}$ that appears in the formula for her proper time will be a small number for major portions of the trip. As a result Andrea's clocks (including the biological clocks in her body) register much less time between A and B than Bernard's clocks do. In spite of the fact that 25 years pass on earth, Andrea's clocks will measure a proper time of (approximately) $\sqrt{1-(0.64)^2}\ (25\ \text{y}) \approx$ 19 years between her departure and arrival [we would need to do the integral of Eq. (5.14) more carefully to get a more accurate answer]. So it is indeed a young Andrea of about 44 that bounds out of the spacecraft to greet her substantially older twin Bernard.

But why is Andrea's line of reasoning wrong? And what of the principle of relativity? The answer to both questions is the same: Andrea is in a *noninertial* reference frame. Every time the spacecraft engines fire, first-law detectors in Andrea's frame register a violation of that law (equivalent detectors in the sun's frame would read nothing). It is Andrea who is pressed into her chair as the engines accelerate and decelerate the spacecraft, not Bernard who is sitting at home. Since Andrea's frame is not inertial, the principle of relativity does not apply to her. This takes care of the argument in favor of the *equality* of the twins.

Andrea's use of the proper time formula leads her astray. She should be computing proper times between the specific events A and B using speed and time measurements made in an *inertial* reference frame, which her frame is clearly not.

Thus there is no paradox. Bernard's answer happens to be right (though his reasoning is a bit wrong), basically because he is in a (nearly) inertial frame (and can thus legally apply the proper time formula to the times measured between event A and B) and Andrea is not. Andrea *really* is younger than Bernard.

This situation should be no more paradoxical than the following (more familiar) situation. Imagine that Cathy and Dave set off from a point A on the surface of the earth. Cathy takes a straight path from starting point A to destination B, while Dave takes a curved path. Should they be shocked when they arrive at the destination and find that Dave has walked more miles than Cathy? Hardly! Dave might try to claim that it was Cathy who departed from him and then returned and so must have taken the curved path, but this claim is misleading. The curvature of Dave's path is an absolute physical property of that path, a property that can be displayed in any fixed coordinate system. Dave's personal coordinate system, whose axes change direction every time he takes a new turn, is not appropriate for determining a path's curvature. We do not have trouble accepting the nonequivalence of the distance measured along Dave's curved and Cathy's straight footpaths in *this* case: we should not have trouble accepting the nonequivalence of the proper times measured along Andrea's and Bernard's worldlines in the first case.

5.9 EXAMPLE PROBLEMS

5.9.1 Walking around the Block

Problem You and a friend stand at a street corner and synchronize your watches. You leave your friend (call this event A) and walk around the block, traveling at a constant speed of 2 m/s (about 4.5 mi/h). After a time $\Delta t_{AB} = 550$ s, as measured by your friend's watch, you return (call this event B). How much *less* than 550 s does your watch register between events A and B?

Solution Take your friend's frame to be inertial; your friend's watch thus measures the coordinate time between A and B in that frame (also the spacetime interval!). In your friend's frame, you have a constant speed of $v = 2$ m/s, or in SR units, $v \approx 6.7\times10^{-9}$. This is extremely small compared to 1, so we can employ the approximation discussed in Sec. 5.3. The proper time that you measure between events A and B is therefore given by

$$\Delta\tau_{AB} = \sqrt{1 - v^2}\,\Delta t_{AB} \approx \left(1 - \tfrac{1}{2}v^2\right)\Delta t_{AB} = \Delta t_{AB} - \tfrac{1}{2}v^2\,\Delta t_{AB} \tag{5.17}$$

The *difference* between $\Delta\tau_{AB}$ (the time *you* read between events A and B) and Δt_{AB} (the time your *friend* reads) is thus approximately

$$-\tfrac{1}{2}v^2\,\Delta t_{AB} = -\tfrac{1}{2}(6.7\times10^{-9})^2(550\text{ s}) \approx -2.5\times10^{-14}\text{ s} \tag{5.18}$$

meaning that the time that you measure is about 549.999999999999975 s. Clearly, the difference between this and 550 s is not going to be even remotely measurable. The difference between proper time, coordinate time, and the spacetime interval are all utterly negligible when the clocks measuring proper time or spacetime interval move between the events in question at ordinary velocities. It is no wonder that we all intuitively have the idea that time is universal and absolute!

5.9.2 A Whirling Clock

Problem Imagine that you are at rest in an inertial frame (the Home Frame) and you are whirling a clock around your head at a constant rate on the end of a 3.0-m-long string. A friend compares the reading of the whirling clock as it speeds by with readings from a stationary clock. Find out how long it takes the whirling clock to go once around its circular path if its reading for one cycle is 0.1 percent smaller than the period read by the stationary clock.

Solution The first step is to rephrase the problem in terms of events. Let event A be the event of the whirling clock passing by the stationary clock. Let event B be the next such passage event. The whirling clock measures a proper time between these events. The stationary clock measures a different proper time between the events (because it is also present at both events). Since the stationary clock is at rest in the Home Frame, the time that it reads is the same as coordinate time in the Home Frame. The whirling clock, on the other hand, is noninertial (its velocity is constantly changing direction as it goes around the circle). But since its *speed* is constant, the proper time that it measures between events A and B can also be calculated using Eq. (5.13). We are given that the result is 99.9 percent of the time measured by the stationary clock, so

$$\Delta\tau_{AB,\text{whirl}} = \sqrt{1 - v^2}\,\Delta t_{AB} = 0.999\Delta t_{AB} \tag{5.19}$$

implying that $\sqrt{1 - v^2} = 0.999$ or

$$1 - v^2 = (0.999)^2 \Rightarrow v^2 = 1 - (0.999)^2 \Rightarrow v = \sqrt{1 - (0.999)^2} \approx 0.45 \tag{5.20}$$

The radius of the circle in SR units is $(3.0\text{ m})(1\text{ s}/3.00 \times 10^8\text{ m}) \approx 1.0 \times 10^{-8}\text{ s} \approx 10\text{ ns}$. The coordinate time that a clock traveling at $v \approx 0.045$ would take to go once around this circle is

$$\Delta t_{AB} = \frac{2\pi R}{v} = \frac{2\pi(1.0 \times 10^{-8}\text{ s})}{0.045} \approx 1.40\times10^{-6}\text{ s} \tag{5.21}$$

This implies a frequency of revolution of $1/\Delta t_{AB} \approx 710{,}000$ Hz (!).[1]

[1]This answer makes it clear that the scenario presented in this problem is completely unrealistic. If the whirling clock has a mass of 100 g, the tension supplied by the string has to be roughly 6×10^{13} N, which is 0.7 billion tons. No realistic "string" could supply such a force, even if you could whirl it at such an outrageous frequency.

5.10 THE DOPPLER SHIFT FORMULA (OPTIONAL)

One of the most important applications of Eq. (5.3) is the computation of the shift in the wavelength of light emitted by a "moving source." You may already know that light emitted by a source moving with respect to some inertial frame is observed in that frame to be redshifted (if the object is departing from that frame) or blueshifted (if the object is approaching that frame). It is our purpose in this section to compute the magnitude of this frequency shift. To simplify matters, we will consider only sources moving directly toward or away from the observer in question.

Let us first consider an inertial clock that emits brief flashes of light. Since the clock is inertial and present at each flash event, it measures the spacetime interval Δs between those flashes. The flashes from this clock are received by an observer at rest at the spatial origin of a given inertial reference frame (which we can call the Home Frame). At a given time, this clock is observed in that frame to be somewhere on the positive side of the x axis and moving with a speed v in the $+x$ direction so that the clock is moving *away* from the observer at the origin. This means that the x component of the clock's velocity is $v_x = +v$. What is the time Δt_r between the reception of these flashes by the observer at the Home Frame origin?[1]

Let the emission of any one flash be event A and the emission of the next flash be event B. If the clock is inertial (*or* if the time between flashes is so short that the clock has an essentially constant velocity between flashes), we can use Eq. (5.3) to compute the coordinate time Δt_{AB} between the flashes measured in the Home Frame in terms of the spacetime interval Δs_{AB} measured between the flashes by the flashing clock itself:

$$\Delta s_{AB} = \sqrt{1 - v_x^2}\,\Delta t_{AB} \qquad \text{implying that} \qquad \Delta t_{AB} = \frac{\Delta s_{AB}}{\sqrt{1 - v_x^2}} \tag{5.22}$$

where I have substituted v_x^2 for v^2, since we are dealing with one-dimensional motion.

But this is only part of the story. The time Δt_r between the *reception* of the flashes from these events by an observer in the Home Frame is *not* equal to Δt_{AB} (the time between their *emission* events): we need to worry about the time that it takes the light to travel from the emission events A and B to the observer. If the distance between the clock and the observer remained constant, the light-travel time for each flash would be the same and Δt_r *would* be equal to the coordinate time Δt_{AB} between the events. But in the time Δt_{AB} between the first and second events, the clock moves an x displacement $v_x\,\Delta t_{AB}$ away from the observer, as measured in the Home Frame (see the spacetime diagram shown in Fig. 5.5). Since light moves with the speed of 1 (in SR units) in all inertial frames, it therefore takes light a time $v_x\,\Delta t_{AB}$ *longer* to travel between event B and the observer than it takes to travel between event A and the observer. Thus the time Δt_r between the reception of the light flashes is equal to the coordinate time between the flashes *plus* the extra light-travel time $v_x\,\Delta t_{AB}$ that the second flash takes to get to the origin:

[1] The method used here follows an approach suggested as an exercise in E. F. Taylor and J. A. Wheeler, *Spacetime Physics,* San Francisco: Freeman, 1966, pp. 62–63.

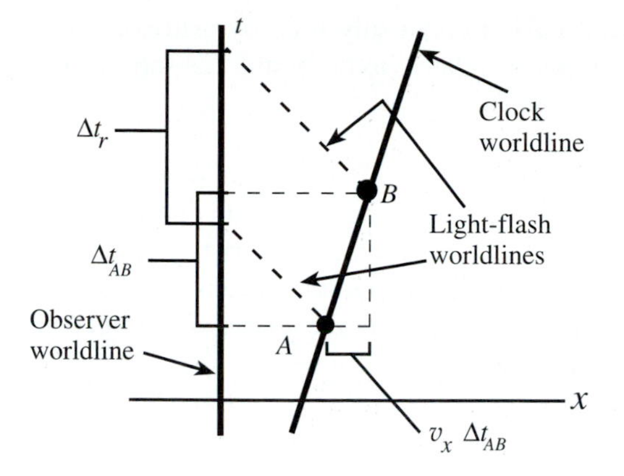

FIGURE 5.5
The time between reception of the flashes Δt_r is not the same as the coordinate time Δt_{AB} between the flashes because the light from flash *B* has to travel an extra distance $v_x\, \Delta t_{AB}$ back to the observer.

$$\Delta t_r = \Delta t_{AB} + v_x\, \Delta t_{AB} = (1 + v_x)\, \Delta t_{AB} \tag{5.23}$$

Using Eq. (5.22) to eliminate Δt_{AB} and using the fact that $1 - v_x^2 = (1 + v_x)(1 - v_x)$, we get

$$\Delta t_r = \frac{(1+v_x)\, \Delta s_{AB}}{\sqrt{1-v_x^2}} = \frac{(1+v_x)\, \Delta s_{AB}}{\sqrt{(1+v_x)(1-v_x)}} = \sqrt{\frac{1+v_x}{1-v_x}}\, \Delta s_{AB} \tag{5.24}$$

Now, light of a definite wavelength can be thought of as being a sequence of electromagnetic pulses separated by a certain characteristic proper time Δs, and so is analogous to the "flashes" considered in the derivation above. Imagine that in a certain inertial frame the time between successive pulses or wavecrests is Δt. The wavelength λ of such a light wave is the distance that a crest will travel in this time, which is equal to $c\, \Delta t = \Delta t$ (in SR units, $c = 1$). So in Eq. (5.24) the wavelength of the emitted light in the source's frame is $\lambda_0 = \Delta s_{AB}$, whereas the wavelength of light as measured by the observer in the Home Frame is $\lambda_r = \Delta t_r$. Plugging these results back into Eq. (5.24) (and dividing through by λ_0), we get

$$\frac{\lambda_r}{\lambda_0} = \sqrt{\frac{1+v_x}{1-v_x}} \tag{5.25}$$

This equation relates the observed wavelength λ_r of light emitted by a moving source to its wavelength λ_0 in the frame of the source itself. Note that if $v_x > 0$ (the source is moving away from the observer), then $\lambda_r > \lambda_0$: the received light has been *redshifted.* On the other hand, if $v_x < 0$ (the source is moving *toward* the observer), $\lambda_r < \lambda_0$: the received light has been *blueshifted.*

Equation (5.25) is called the **Doppler shift formula,** and it applies to any source that moves either directly toward or directly away from an observer (a more complicated formula is required if the source moves tangentially as well as radially with respect to the observer). Since excited atoms often emit light having a discrete and characteristic set of wavelengths, this formula is commonly used by astronomers to compute the radial speed of astronomical objects relative to the earth. The formula has a variety of other scientific applications as well.

Example Problem Light from excited atoms in a certain quasar is received by observers on the earth. The wavelength of a certain spectral line of this light is measured by these observers to be $\lambda_r = 1.12\lambda_0$, where λ_0 is the characteristic wavelength of the spectral line measured in the rest frame of the atoms (this means the light has been *redshifted* by about 12 percent). What is the velocity of the quasar with respect to the earth (assuming that it is traveling directly away)?

Solution From Eq. (5.25), we have

$$1 + \sqrt{\frac{1 + v_x}{1 - v_x}} = \frac{\lambda_r}{\lambda_0} = 1.12 \equiv u \tag{5.26}$$

where I have given the ratio the name u for convenience. Squaring both sides and solving for v_x, we get

$$\begin{aligned} u^2 &= \frac{1 + v_x}{1 - v_x} \\ &\Rightarrow (1 - v_x)u^2 = 1 + v_x \\ &\Rightarrow u^2 - v_x u^2 = 1 + v_x \\ &\Rightarrow u^2 - 1 = v_x(1 + u^2) \\ &\Rightarrow v_x = \frac{u^2 - 1}{u^2 + 1} = \frac{(1.12)^2 - 1}{(1.12)^2 + 1} = 0.11 \end{aligned} \tag{5.27}$$

The speed of the quasar is thus about 11 percent of the speed of light.

5.11 SUMMARY

Equation (5.3) is an elementary consequence of the metric equation that has a wide range of applications in relativistic physics. This equation tells us that

$$\Delta s = \sqrt{1 - v^2}\, \Delta t \tag{5.3}$$

In other words, the spacetime interval Δs between two events (i.e., the proper time measured by an inertial clock present at both events) is *smaller* than the coordinate time interval Δt between the *same* events (as measured in some inertial reference frame) by the factor $\sqrt{1 - v^2}$, where v is the speed of the clock measuring Δs relative to the same frame where Δt is measured.

Note that when $v << 1$, $\Delta s \approx \Delta t$, meaning that we do not have to worry about the distinction between coordinate time and the spacetime interval in this limit (this is the normal situation in daily life). On the other hand, as $v \to 1$, Δs becomes smaller and smaller compared to Δt. If $v > 1$, Δs actually becomes *imaginary* (not negative!), signaling that we have violated the assumptions used to derive the metric equation in Sec. 4.2.

The **binomial approximation,**

$$(1+x)^a \approx 1 + ax \qquad \text{if } |x| << 1 \tag{5.4}$$

is very useful for computing the factor $\sqrt{1 - v^2}$ when v is small (as in daily life):

$$\sqrt{1 - v^2} \approx 1 + \frac{1}{2}(-v^2) = 1 - \frac{1}{2}v^2 \qquad \text{if } v << 1 \tag{5.5}$$

"Small" in this case depends on how accurate you wish to be, but this approximation is accurate to at least three decimal places in the *difference* between 1 and $\sqrt{1 - v^2}$ when $v < 0.1$.

Equation (5.3) can also be applied to the case of the proper time $\Delta\tau$ between two events measured by an accelerating clock by dividing up the clock's worldline into small enough segments so that each segment is essentially straight; i.e., the duration of time covered by each segment is so short that the clock's speed is essentially constant during the interval. (This is essentially the same method that would be used to compute the length of a curved path on a plane, and thus represents another application of the geometric analogy between proper time and pathlength.) In the limit that the coordinate time between events becomes infinitesimal, we thus have

$$d\tau = \sqrt{1 - v^2}\, dt \tag{5.11}$$

To find the finite proper time between two events measured by a clock traveling along a certain worldline, one simply integrates Eq. (5.11) over the worldline:

$$\Delta\tau_{AB} = \int_{\text{worldline}} \sqrt{1 - v^2}\, dt \tag{5.12}$$

If the speed v (not the velocity) of the accelerating clock is constant along its worldline, the integral in Eq. (5.12) is easy, and we have

$$\Delta\tau_{AB} = \sqrt{1 - v^2}\, \Delta t_{AB} \qquad \text{if } v = \text{constant} \tag{5.13}$$

Equations (5.3) and (5.12) imply the following strict relationship between the coordinate time Δt, the spacetime interval Δs, and any proper time $\Delta\tau$ measured between a *given* pair of events:

$$\Delta t \geq \Delta s \geq \Delta\tau \tag{5.14}$$

The first inequality is an equality when the events occur at the same place (i.e., same clock) in the inertial frame where Δt is measured, and the last inequality is an equality if the clock measuring the proper time $\Delta\tau$ is an *inertial* clock. This inequality also implies that a straight (i.e., constant-velocity) worldline between two events is the worldline along which one would measure the *longest* possible proper time between those events.

We can use Eq. (5.3) and the constancy of the speed of light to derive the **Doppler shift formula**

$$\frac{\lambda_r}{\lambda_0} = \sqrt{\frac{1 + v_x}{1 - v_x}} \tag{5.25}$$

This equation describes how the wavelength λ_r of light received by an observer at the spatial origin of an inertial reference frame is related to the wavelength λ_0 of light emitted by a source moving along the positive side of the x axis in that frame with x velocity v_x. When $v_x > 0$ (the source is moving away from the observer), the light is redshifted ($\lambda_r > \lambda_0$), but when $v_x < 0$ (the source is moving toward the observer), the light is blueshifted ($\lambda_r < \lambda_0$).

PROBLEMS

5.1 RELATIVISTIC TRAVEL. Say you were traveling at a constant velocity from the earth to the center of the Milky Way galaxy (which is 30,000 light-years away). What constant speed do you need to have to cover this distance within your remaining lifetime ($\approx$ 50 years, as measured by your watch and internal biological clocks). Be sure in your solution to describe the events that you are talking about when you do this calculation.

5.2 RELATIVISTIC TRAVEL. A muon travels from where it is created in the upper atmosphere toward the earth's surface at a speed of 0.9925. Say that the muon survives for 2.22 μs (as measured by its interior clock) before it decays. How long does it last according to clocks on the earth? How far does it travel in the earth's frame?

5.3 Verify that the centripetal acceleration of the muons in the experiment described in Sec. 5.7 is indeed about $10^{15}g$. The ring radius in this experiment was 14.02 m.

5.4 PROPER TIME AROUND THE WORLD. Clock P is synchronized with clock Q. Clock P is then put on a jet plane that travels around the world in 30.2 h. Clock P is then taken off the plane and compared again with Q. Calculate the difference between the clocks' final readings. Express your answer in nanoseconds. (This is a simplified version of the Halefe-Keating experiment described in Sec. 5.7.)

5.5 JET LAG. A new hyperjet is constructed that can go all the way around the world in 6.235 s as measured by clocks at the airport. Assuming that the jet cruises at a constant speed, how long do the pilots' watches register for a complete circumnavigation of the globe? The radius of the earth is about 6380 km.

5.6 HOW GOOD IS THE APPROXIMATION? Compare both sides of Eq. (5.5) for the following values of v: 0.5, 0.2, 0.1, 0.05, 0.01, 0.005.

5.7 JOGGERS AGE MORE SLOWLY. A jogger runs exactly 22 times around a 1/2-km track in 48 min, according to a friend sitting at rest on the side. If the jogger and friend synchronized watches before the run, how much are they out of synchronization afterward?

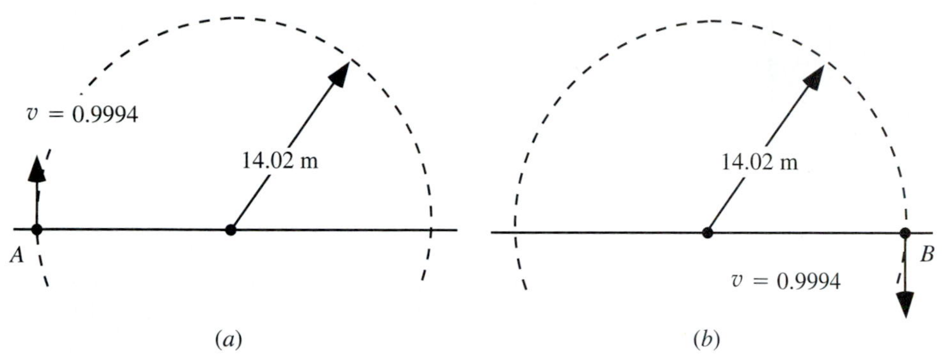

FIGURE 5.6
(*a*) The particle at event *A*. (*b*) The particle at event *B*.

5.8 THE THREE KINDS OF TIME. The designers of particle accelerators use electromagnetic fields to boost particles to relativistic speeds while at the same time constraining them to move in a circular path inside a donut-shaped evacuated cavity. Imagine a particle traveling in such an accelerator in a circular path of radius 14.02 m at a constant speed of 0.9994 (as measured by laboratory observers). Let event *A* be the particle passing a certain point on its circular path, and let event *B* be the particle passing the point of the circle directly opposite that point, as illustrated in Fig. 5.6.

a. What is the coordinate time Δt and the distance Δd between these events in the laboratory frame? [*Hint:* $\Delta d \neq \pi(14.02\text{ m})$! Think about it more carefully!]

b. What is the spacetime interval Δs between the events?

c. What is the proper time $\Delta\tau$ between the events as measured by a clock traveling with the particle? About how many times greater than $\Delta\tau$ is Δt?

5.9 MUON STORAGE. The half-life of a muon at rest is 1.52 μs. One can store muons for a much longer time (as measured in the laboratory) by accelerating them to a speed very close to that of light and then keeping them circulating at that speed in an evacuated ring. Assume that you want to design a ring that can keep muons moving so fast that they have a laboratory half-life of 0.25 s (about the time it takes a person to blink). How fast will the muons have to be moving? If the ring is 14.02 m in diameter, how long will it take a muon to circle the ring (as measured in the laboratory)? (*Hint:* Define $u \equiv \Delta s/\Delta t$ and use the approximation given by Eq. 5.4 to help you answer the first question.)

5.10 THE THREE KINDS OF TIME. The time between two events *A* and *B* is measured by a clock following worldline *P*, by a clock following worldline *Q*, and by reading the difference in the event times registered by the synchronized pair of clocks following worldlines *R* and *S*, as shown in Fig. 5.7.

a. Which clock or clocks measure proper time between *A* and *B*?

b. Which clocks (if any) measure the spacetime interval between these events?

c. Which of the three measurements described above yields the largest result? Which yields the shortest? Support your answer with a short argument.

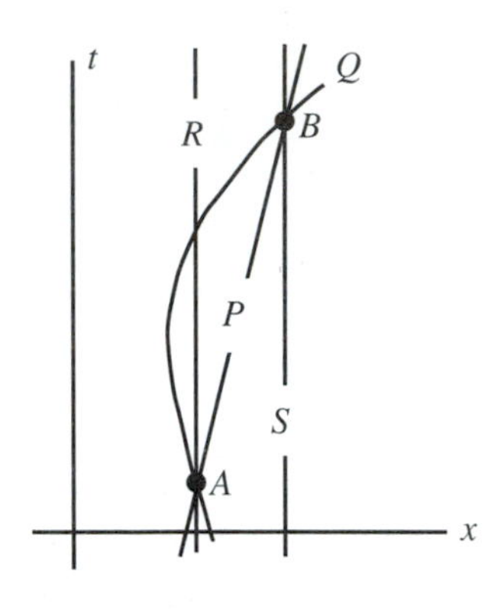

FIGURE 5.7
The time between events A and B is measured by clock following worldline P, a clock following Q, and by finding the difference of the readings of clocks following worldlines R and S at the respective events.

5.11 THE RELATIVE MAGNITUDES OF TIME INTERVALS. In Sec. 5.6, it was proved that the coordinate time, the spacetime interval, and the proper time between a given pair of events stand in the relationship $\Delta t \geq \Delta s \geq \Delta\tau$, no matter what inertial frame is used to measure Δt and no matter what worldline is followed by the clock measuring $\Delta\tau$. It was *asserted* there that $\Delta t = \Delta s$ if and only if Δt is measured in a frame where the distance Δd between these events is zero. It was also *asserted* that $\Delta s = \Delta\tau$ if and only if the clock measuring $\Delta\tau$ is inertial. Write a short argument supporting each of these statements (for both directions of the "if and only if").

5.12 PROOF OF THE BINOMIAL APPROXIMATION. It is possible to prove the binomial approximation as follows. Consider the function $f(x) = 1 + x$. According to the definition of the derivative, we have [for *any* function $f(x)$, actually]

$$\frac{f(x) - f(0)}{x} \approx \left[\frac{df}{dx}\right]_{x=0} \quad \text{if } x \text{ is very small} \tag{5.28}$$

where $[df/dx]_{x=0}$ tells you to evaluate the derivative at $x = 0$. Use this expression and the chain rule to arrive at Eq. (5.4).

5.13 PROPER TIME FOR A VARYING SPEED. Consider an inertial frame at rest with respect to the earth. An alien spaceship is observed to move along the x axis of this frame with $x(t) = [\sin(\omega t + \pi/5) - b]/\omega$, where both x and t are measured in the given frame, $\omega = (\pi/2)\, h^{-1}$, and $b = \sin(\pi/5)$. Assume that the earth is located at $x = 0$ in this frame.

a. Argue that the ship passes the earth at $t = 0$ and again at $t = 1.0$ h. (*Hint:* Note that $\omega t = \pi/2$ at this time.)

b. Draw a quantitatively accurate spacetime diagram of the spaceship's worldline, labeling the events where and when it passes the earth as A and B.

c. Show that the x velocity of the ship is $v_x = \cos(\omega t + \pi/4)$ as measured in the earth frame. (*Hint:* No knowledge of relativity needed!)

d. Find the proper time measured by clocks on the ship between the time the ship passes earth the first time and the time it passes earth the second time. (*Hint:* $1 - \cos^2 x = \sin^2 x$. You can look up any integral you might need.)

5.14 DOPPLER SHIFT. A spaceship moves directly away from earth at a speed of 0.5. By what factor is the spaceship's taillight redshifted?

5.15 DOPPLER SHIFT. The hydrogen fusion flare from the engine of an alien ship is detected from earth. If the light from the hydrogen atoms is observed to be blueshifted by 35 percent compared to its normal wavelengths, how rapidly in the alien ship approaching earth?

5.16 SPEEDING.[1] A physicist is brought to court for running a red light. The physicist argues in court that the car was traveling so fast toward the intersection that the light looked green. The judge changes the charge to speeding and fines the physicist a tenth of a penny for every mile per hour the physicist was traveling over the local speed limit of 45 mi/h. What was the fine (approximately)? (*Useful information:* $\lambda_{red} \approx 650$ nm $= 650 \times 10^{-9}$ m, $\lambda_{green} \approx 530$ nm, 2.2 mi/h $\approx$ 1.0 m/s. Note that the light is *blue*shifted.)

[1]Adapted from E. F. Taylor and J. A. Wheeler, *Spacetime Physics,* San Francisco: Freeman, 1966, p. 156.

6

COORDINATE TRANSFORMATIONS

From this hour on, space as such and time as such shall recede into the shadows and only a kind of union of the two retain significance.

H. Minkowski[1]

6.1 INTRODUCTION

In Chaps. 4 and 5, we explored consequences of the metric equation that links the spacetime interval Δs between two events to coordinate differences between those events measured in any inertial frame. The extension of the metric equation to curved worldlines [Eq. (5.12)] allowed us to compute the proper time $\Delta\tau$ along a given worldline using measurements found in an inertial frame. In short, our focus in the last two chapters was to link coordinate measurements (which are frame-*dependent*) to the frame-*in*dependent quantities Δs and $\Delta\tau$.

To delve further into the implications of the principle of relativity, we need to go a step further: we need to find a way of linking an event's t, x, y, z coordinates measured in one inertial frame with the same event's t', x', y', z' coordinates measured in another inertial frame. These equations that link coordinates in one frame with those in another are called the **Lorentz transformation equations** and are the relativistic generalization of the galilean transformation equations [Eqs. (1.2)]. We need these equations to find the relativistic generalization of the galilean velocity transformation equa-

[1]Lorentz, A. Einstein, et al., *The Principle of Relativity,* New York: Dover, 1952, p. 75.

tions [Eqs. (1.3)], to explore the phenomenon of "length contraction," to explain why nothing can go faster than the speed of light, and so on.

The derivation of the Lorentz transformation equations is direct but somewhat abstract, and that abstraction can blunt one's intuition about what is really going on. Therefore, in this text we will address the same problem using a more visual, intuitive tool called a **two-observer spacetime diagram.** In a two-observer spacetime diagram, we superimpose the coordinate axes for two different observers on the same spacetime diagram. What we will end up with will be analogous to the drawing shown in Fig. 6.1, which shows two ordinary cartesian coordinate systems (one rotated with respect to the other) superimposed upon the plane.

Once we have set up the Cartesian two-observer diagram shown in Fig. 6.1, if we know the coordinates of a point A in the xy coordinate system (shown in Fig. 6.1 as having coordinates $x_A = 7$ m, $y_A = 4$ m), we can plot point A relative to point O. Then we can just *read* the coordinates of A in the $x'y'$ coordinate system from the diagram (the coordinates are $x'_A = 8$ m, $y'_A = 1$ m in this case). We do not have to use any equations or do any calculations at all!

Setting up such a diagram is easy enough for plane cartesian coordinate systems: it merely involves drawing two sets of perpendicular axes (one rotated with respect to the other), scaling the axes, and drawing coordinate grid lines for each set of axes. Setting up the two sets of coordinate axes representing different inertial frames on a spacetime diagram is a *similar* process, but the peculiarities of spacetime geometry relative to plane geometry lead to some surprising dissimilarities as well. Therefore we need to develop the procedure in a careful step-by-step manner so that we are sure to catch all these dissimilarities.

FIGURE 6.1
A drawing showing two sets of cartesian coordinate axes superimposed on the same plane. The coordinates of point A in both coordinate systems can be read easily from such a diagram.

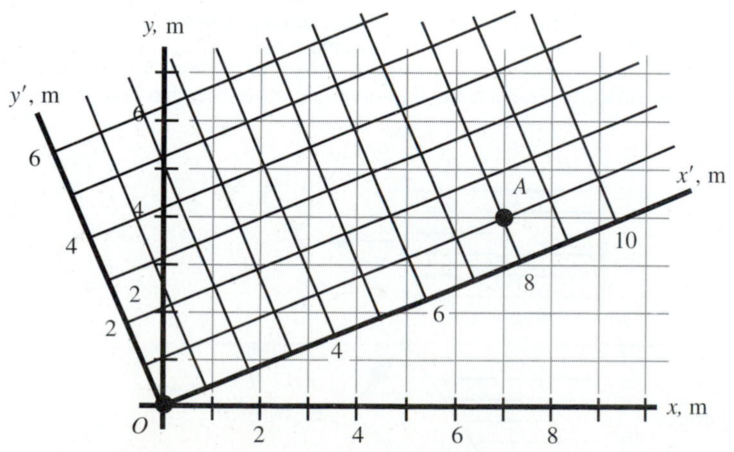

6.2 CONVENTIONS

To make the task of constructing two-observer spacetime diagrams easier, it is convenient to make several assumptions. First, *we assume that the two inertial reference frames are in standard orientation* with respect to each other, as defined in Sect. 1.9. That is, we assume that their corresponding spatial axes point in the same directions in space and the relative motion of the frames is directed along the common x direction. Since one can choose whatever orientation desired for the axes of a spatial coordinate system, we do not really lose anything by choosing the frames to have this orientation, and we gain much in simplicity.

Second, *we will work with only those events that occur along the common x and x' axes of these frames* (i.e., those having coordinates $y = y' = z = z' = 0$). This is a substantial concession to convenience: we would really like to be able to handle any event. But plotting an event with arbitrary coordinates on a spacetime diagram would require that the diagram have four dimensions, which is impossible to represent on a sheet of paper. We choose therefore to limit our attention to events that can be easily plotted on a two-dimensional spacetime diagram. We will see that many interesting problems can still be treated within this restriction. So until you are told otherwise, you should assume that $y' = y = z' = z = 0$ for all events under discussion.

The first step in actually drawing a two-observer spacetime diagram is to pick one of the two frames to be the **Home Frame.** The other frame can be called the **Other Frame** (easy to remember, right?). Remember that the terms *Home Frame* and *Other Frame* are capitalized in this text to emphasize that these phrases are actually *names* of inertial frames.

It is conventional (but not absolutely necessary) to pick the Other Frame to be the frame of the two that moves in the $+x$ direction with respect to the Home Frame, as shown in Fig. 6.2 (alternatively, one can think of the Home Frame as the one that moves in the $-x$ direction with respect to the Other Frame): some signs in equations

FIGURE 6.2
Two inertial reference frames in standard orientation. The frames are represented schematically here by bare orthogonal axes. Note that when frames are in standard orientation, the Other Frame is always taken to be the frame moving in the *positive* direction along the common *x* direction (alternatively, we can say that the Home Frame is the frame moving in the negative *x* direction).

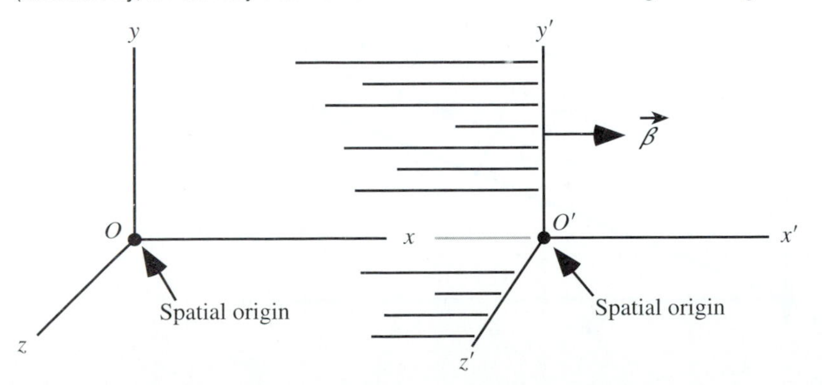

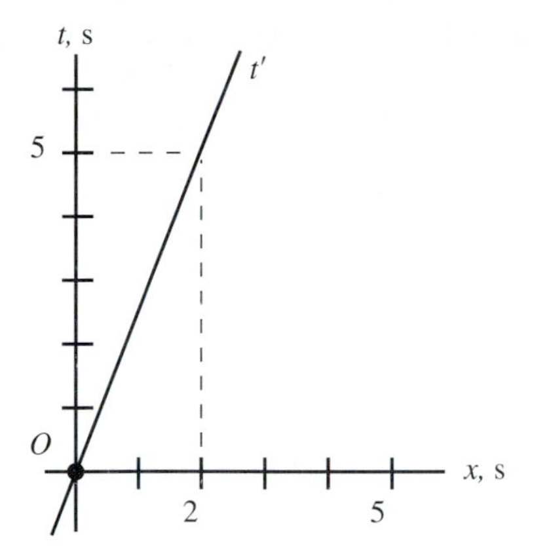

FIGURE 6.3
The time axes for both the Home and Other frames. The t axis connects all events occurring at $x = 0$ in the Home Frame, while the t' axis connects all events occurring at $x' = 0$ in the Other Frame. The diagram is drawn assuming that the Other Frame is traveling with a speed $\beta = 2/5$ with respect to the Home Frame: therefore the slope of the t' axis is $1/\beta = 5/2$.

that follow assume we are following this convention. It is also conventional to distinguish each frame's axes and coordinate measurements by using t, x, y, z for Home Frame axes and coordinates and t', x', y', and z' for the Other Frame axes and coordinates.

Now, our choice of a Home Frame does not *necessarily* mean we are considering that frame to be at rest and the Other Frame to be moving: we still want to reserve the freedom to consider either frame as being at rest. What this choice does imply is simply that we will represent the Home Frame t and x axes in the *usual* manner in a spacetime diagram (i.e., its t axis will be vertical and its x axis will be horizontal).

6.3 DRAWING THE TIME AXIS FOR THE OTHER FRAME

The **time axis** for any frame on a spacetime diagram is by definition the line connecting all events that have x coordinate $= 0$ in that frame. This means that the time axis is the worldline of the clock at the spatial origin of the reference frame (all events happening at the spatial origin of a reference frame have spatial coordinate $x = 0$ by definition). The t axis of the Home Frame is drawn as a vertical line by convention. How should the t' axis of the Other Frame be drawn?

The Other Frame moves with speed β along the $+x$ direction with respect to the Home Frame by hypothesis. This means that as measured in the Home Frame, the Other Frame's spatial origin moves β units in the $+x$ direction every unit of time. The worldline of the Other Frame's origin as plotted on the spacetime diagram is thus a straight line of slope $1/\beta$ as shown in Fig. 6.3. Note that this line goes through the origin event O since the spatial origins of both frames coincide at $t = t' = 0$ if the frames are in standard orientation.

6.4 CALIBRATING THE t' AXIS

The next step is to put an appropriate scale on the Other Frame's t' axis. It is (unfortunately) not correct to simply mark this axis using the same scale as used for the t and x axes. The purpose of this section is to describe how to correctly mark a scale on the t' axis.

Consider the marks on the t' axis labeled A and B on Fig. 6.4. Each of these marks corresponds to an event in spacetime: in the Other Frame, these events are separated by the coordinate differences $\Delta t'_{AB} = 1$ s and $\Delta x'_{AB} = 0$ (since *all* events along the t' axis occur at the spatial origin of the Other Frame by definition). Let Δt_{AB} and Δx_{AB} be the coordinate differences between the same two events as measured in the Home Frame. Since the spacetime interval between these two events is frame-independent, we have

$$(\Delta t_{AB})^2 - (\Delta x_{AB})^2 = (\Delta s_{AB})^2 = (\Delta t'_{AB})^2 - (\Delta x'_{AB})^2 = (\Delta t'_{AB})^2 \tag{6.1}$$

because $\Delta x'_{AB} = 0$ in this case. But in the previous section, we decided that the slope of the t' axis was $1/\beta$ (where β is the velocity of the Other Frame with respect to the Home Frame). This means that Δt_{AB} and Δx_{AB} must be related as follows:

$$\frac{\Delta x_{AB}}{\Delta t_{AB}} = \beta \qquad \text{or} \qquad \Delta x_{AB} = \beta\, \Delta t_{AB} \tag{6.2}$$

(Another way to think of this is to remember that the t' axis is the worldline of the master clock at the spatial origin of the Other Frame, which in the time Δt_{AB} between the events travels a distance $\Delta x_{AB} = \beta\, \Delta t_{AB}$, since it is moving at speed β by definition.) Plugging Eq. (6.2) into (6.1), we get

$$(\Delta t'_{AB})^2 = (\Delta t_{AB})^2 - (\beta\, \Delta t_{AB})^2 = (\Delta t_{AB})^2(1 - \beta^2) \tag{6.3a}$$

and solving for Δt_{AB}, we get

$$\Delta t_{AB} = \frac{\Delta t'_{AB}}{\sqrt{1 - \beta^2}} \tag{6.4}$$

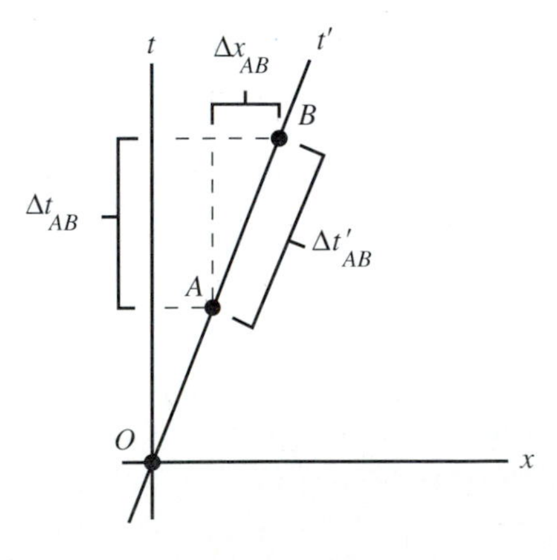

FIGURE 6.4
This figure shows two arbitrary scale marks on the t' axis, separated by a time $\Delta t'_{AB}$. Since the axis represents the worldline of the spatial origin of the Other Frame as it moves to the right with speed β, in the time Δt_{AB} between events A and B, the origin will move a distance $\Delta x_{AB} = \beta\, \Delta t_{AB}$.

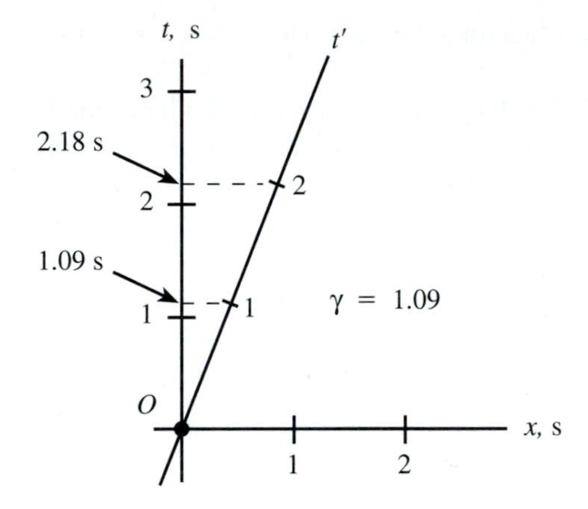

FIGURE 6.5
Calibrating the t' axis of an Other Frame moving at a speed of 2/5 relative to the Home Frame. Marks separated by a time $\Delta t' = 1$ s in the Other Frame are separated by $\Delta t \approx 1.09$ s in the Home Frame. Note that the distance between the marks on the t' axis is "stretched out" compared to the distance between the marks on the t axis.

The quantity $1/\sqrt{1-\beta^2}$ occurs so often in relativity theory that it is given its own special symbol (the Greek letter *gamma*):

$$\gamma \equiv \frac{1}{\sqrt{1-\beta^2}} \tag{6.5}$$

(Note that γ is a number that is always larger than 1.) Using this definition, Eq. (6.4) becomes

$$\Delta t_{AB} = \gamma\, \Delta t'_{AB} \tag{6.6}$$

The point of all this is that if you want to draw marks on the t' axis of the graph that are separated by some time interval $\Delta t'$, then you must draw these marks so that they have a vertical separation of $\Delta t = \gamma\, \Delta t'$. To give a concrete example, imagine that your Other Frame moves at a speed of $\beta = 2/5$ and you want to draw marks on the t' axis for that frame that are separated by $\Delta t' = 1$ s. According to Eq. (6.6), you need to draw these marks so that they have a vertical separation of

$$\Delta t = \gamma\, \Delta t' = \frac{1\text{ s}}{\sqrt{1-(2/5)^2}} \approx 1.09\text{ s} \tag{6.7}$$

as shown in Fig. 6.5.

6.5 DRAWING THE DIAGRAM *x′* AXIS

The t' axis on the spacetime diagram is defined to be the line connecting all events that occur at $x' = 0$ (that is, the spatial origin) of the Other Frame. Analogously, the **diagram *x′* axis** is defined to be the line connecting all events that occur at $t' = 0$ (that is, at the same *time* as the origin event). The diagram x axis for the Home Frame is conventionally drawn as a horizontal line on a spacetime diagram. How should we draw the diagram x' axis of the Other Frame?

The natural thing to do would be to draw the diagram x' axis perpendicular to the t' axis. Unfortunately, this approach is *not* consistent with the definition of the diagram x' axis given above. To figure out the right way to draw this axis, we have to carefully consider the implications of the fact that *the diagram x' axis connects events that are simultaneous in the Other Frame.*

(*Note:* The phrase "diagram x' axis" or "diagram x axis" is going to refer to *a line drawn on a spacetime diagram that connects all events that occur at zero time.* This is to be sharply distinguished from the line in physical *space* that goes through the spatial origin and connects all points having $y = z = 0$: when I want to talk about the latter, I will speak of the "x direction" or the "spatial x axis.")

We begin by considering a set of events that illustrates the use of the radar method to determine coordinates in the Other Frame. Imagine that at time $t' = -T$ (where T is some arbitrary number) a light flash is sent from the master clock (located at the spatial origin of the Other Frame) in the $+x$ direction (call the emission of the flash event A). This flash reflects from some event B and returns to the Other Frame's master clock at time $t' = +T$ (call the reception of the flash by the master clock event C). Since the light flash takes an equal amount of time to return from event B as it took to get there, we conclude that event B happened at a time halfway between $t'_A = -T$ and $t'_C = +T$, that is, $t'_B = 0$. This means that event B will be *simultaneous* with the origin event O. This is illustrated in Fig. 6.6.

Now let us draw this set of events on a spacetime diagram based on the Home Frame (see Fig. 6.7). Events A, O, and C all occur at $x' = 0$, so they all lie on the t' axis of the spacetime diagram. Moreover, events A and C occur at $t' = -T$ and $t' = +T$, so they must be symmetrically spaced on opposite sides of the origin event, as shown. The worldline of the clock present at event B must be parallel to and just far

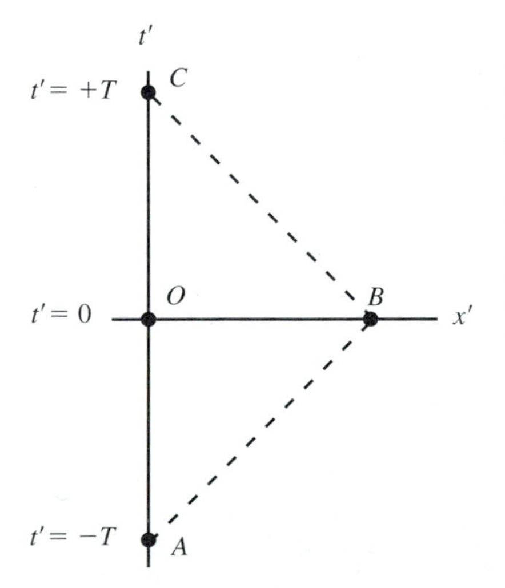

FIGURE 6.6
Spacetime diagram illustrating the use of the radar method to determine the time coordinate of event B. A flash of light is emitted from the spatial origin of the Other Frame at time $t' = -T$ (event A). This flash is reflected at event B and returns to the spatial origin at time $t' = +T$ (event C). Because the light flash travels the same distance down and back in the spatial x' direction, event B must occur halfway between times $-T$ and $+T$; that is, it must occur at time $t' = 0$. Thus event B lies on the diagram x' axis of this spacetime diagram by definition.

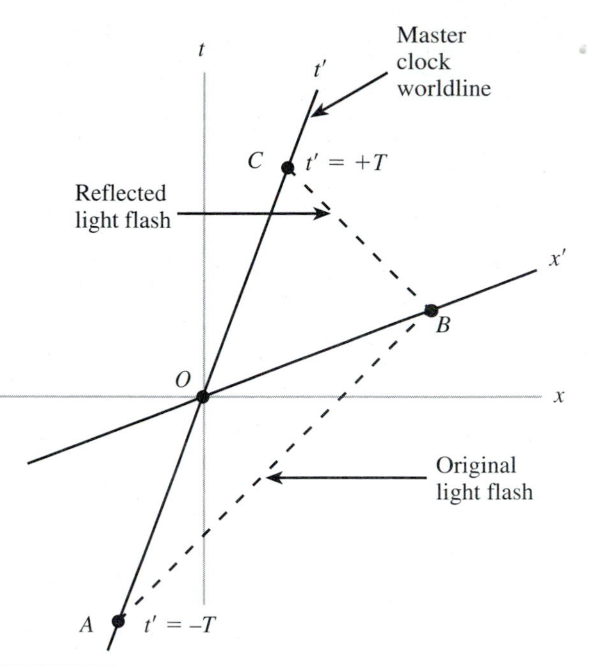

FIGURE 6.7
A spacetime diagram of the set of events illustrated in Fig. 6.6 as plotted by an observer in the Home Frame. Note that *A* and *C* are symmetrically placed on the t' axis on opposite sides of the origin event *O*. Note that both event *O* and event *B* occur at time $t' = 0$ by definition of clock synchronization in the Other Frame. Therefore, these events must lie along the diagram x' axis, implying that the diagram x' axis *must* have an upward slope as shown.

enough away from the t' axis so that the light-flash worldlines for the right- and left-going flashes meet at the worldline of the second clock. Note that since the speed of light is 1 in every reference frame, these light-flash worldlines must have a slope of ±1 on this diagram.

Now, by the definition of clock synchronization in the Other Frame, events *O* and *B* both occur at time $t' = 0$. The diagram x' axis is defined to be the line connecting all events that occur at time $t' = 0$. Therefore, the diagram x' axis must go through events *O* and *B*. This means that *the diagram* x' *axis must angle upward,* as shown in Fig. 6.7.

In fact, by considering the geometric relationships implicit in Fig. 6.7, I can show you that *the diagram* x' *axis makes the same angle with the diagram* x *axis that the* t' *axis makes with the* t *axis.* The argument goes like this. Imagine that the master clock emits a right-going light flash at the origin event *O*, and let event *E* be this flash meeting the incoming reflected flash. (This new light flash has no physical importance: it just makes the following argument simpler.)

Since light-flash worldlines always have a slope of ±1, they all make a 45° angle with respect to the vertical or horizontal directions. This means that if light-flash worldlines cross at all, they always cross at right angles (see Fig. 6.8).

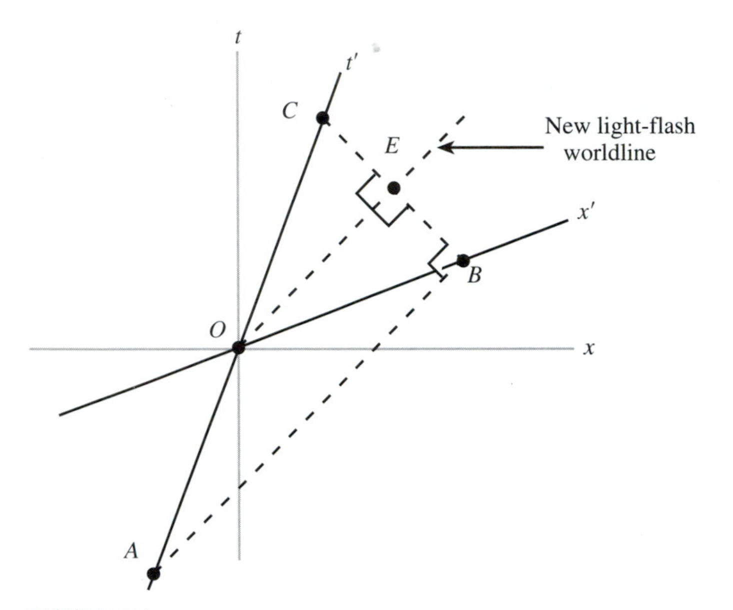

FIGURE 6.8
The spacetime diagram shown in Fig. 6.7 with the addition of a right-going light flash emitted by the master clock at the origin event *O*. Because the light-flash worldlines all have slopes of ±1, they all make an angle of 45° with respect to the vertical. This means that these worldlines always cross at right angles.

Now, I claim that triangles *ABC* and *OEC* are similar triangles: they are right triangles that share the common angle ϕ (see Fig. 6.9*a*). Moreover, the hypotenuse of *ABC* is *twice* as long as that of *OEC,* since *A* and *C* are symmetrically placed about the event *O*. This means that the triangle *ABC* must be exactly twice as large as the triangle *OEC,* implying that the line *BC* must also be twice as large as the line *EC* (remember that if two triangles are similar, the lengths of their corresponding sides are proportional). But if line *BC* is twice as large as *EC,* then the length of line *BE* must be equal to the length of line *EC* (see Fig. 6.9*b*.)

But this means that triangles *OEC* and *OEB* must be *identical,* since they are both right triangles and their corresponding legs are equal in length. This means that $\alpha_1 = \alpha_2$, which in turn means that $\theta_1 \equiv 45° - \alpha_1$ is equal to $\theta_2 \equiv 45° - \alpha_2$. Thus *the diagram x′ axis makes the same angle with the diagram x axis that the t′ axis makes with the t axis,* as previously asserted.

Another important consequence is that the length of the line *OC* (which represents the coordinate time interval $\Delta t' = +T$) is the same as the length of the line *OB* (which represents the coordinate displacement $\Delta x' = T$, the distance that the light signal had to travel to get to event *B*). This means that *the scale of both axes must be the same;* that is, the spacing of marks on the diagram x' axis will be exactly the same as the spacing of marks on the t' axis!

Note that $\tan\theta_1$ = run/rise for the t' axis = 1/slope of t' axis = β. Note also that

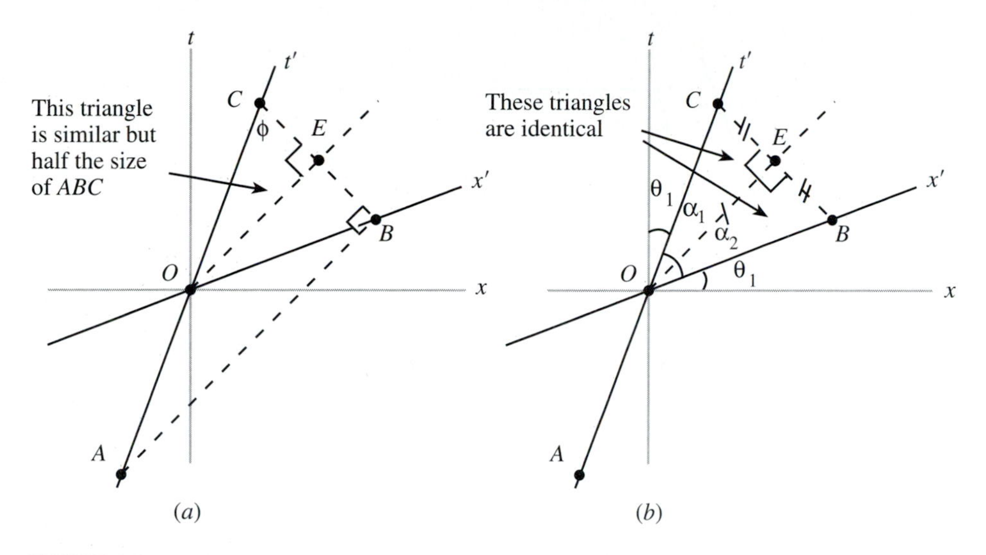

FIGURE 6.9
(*a*) Triangles *ABC* and *OEC* are similar (they are right triangles and share the common angle ϕ). *OEC* is half the size of *ABC,* since line *AO* has the same length as line *OC*. But this means that line *BC* is twice as long as line *EC,* implying that line *BE* has the same length as line *EC*. (*b*) But this implies that triangles *OEC* and *OEB* are identical since they are right triangles whose corresponding legs are equal in length. Therefore the two α angles are equal, implying that the two θ angles are equal. Note also that the length of line *OC* is equal to the length of line *OB.*

$\tan\theta_2$ = rise/run for the diagram x' axis = slope of diagram x' axis. Since we have just seen that $\tan\theta_1 = \tan\theta_2$, we have

$$\text{Slope of } x' \text{ axis} = \beta \tag{6.8}$$

So, to be consistent with the principle of relativity, we must draw the Other Frame t' and diagram x' axes with slopes $1/\beta$ and β, respectively.

6.6 READING THE TWO-OBSERVER DIAGRAM

In summary, what have we discovered? To construct a two-observer spacetime diagram, we first construct the Home Frame's t axis and diagram x axis perpendicular with each other (with the t axis vertical). Then we draw the t axis of the Other Frame with slope $1/\beta$ and the diagram x' axis of the Other Frame with slope β. We then calibrate the Other Frame time axis with marks that are separated vertically by $\Delta t = \gamma\,\Delta t'$ (where $\Delta t'$ is the time interval between the marks in the Other Frame and $\gamma \equiv 1/\sqrt{1-\beta^2}$). Finally, we calibrate the diagram x' axis with marks separated by the *same* distance as marks on the t' axis (i.e., the marks should be separated horizontally by $\Delta x = \gamma\Delta x'$, where $\Delta x'$ is the spatial separation between the marks in the Other Frame).

We can now find the t' and x' coordinates of any event on the diagram as follows. The t' axis is by definition the line on the spacetime diagram connecting all events

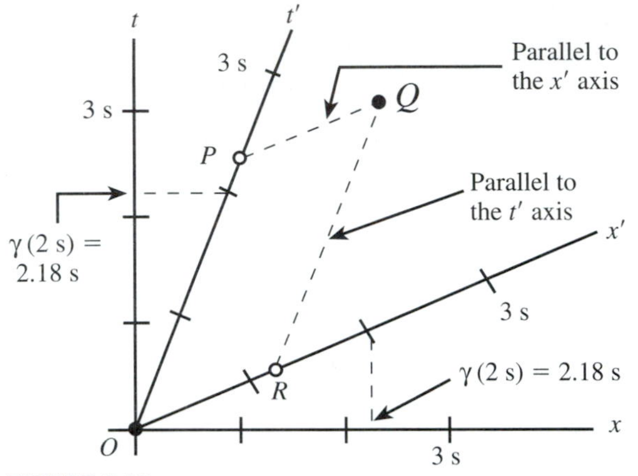

FIGURE 6.10
Events Q and R occur at the same *place* in the Other Frame, since all events that occur at the same place in that frame lie along a line parallel to the t' axis. Similarly, events P and Q occur at the same *time* in the Other Frame. Thus the time of Q is the same as the time of P in that frame (that is, $t'_Q \approx 2.3$ s in this case), and the position of Q is the same as the position of R (that is, $x'_Q \approx 1.2$ s in this case). The relative speed of the frames is $\beta = 2/5$ in the case shown.

that occur at $x' = 0$. The line connecting all events that have coordinate $x' = 1$ s (or any given value not equal to zero) will be a line *parallel* to the t' axis, because the Other Frame's lattice clock at $x' = 1$ s moves at the same velocity as the master clock at $x' = 0$, and the latter's worldline defines the t' axis.

Similarly, the line on the diagram connecting all events that have the same t' coordinate must be a line parallel to the diagram x' axis (the line connecting all events having $t' = 0$). Here is an argument for this statement: if the line connecting all events that occur at $t' = 1$ s, for example, were *not* parallel to the line connecting all events that occur at $t' = 0$, then these lines would intersect at some point on the diagram. The event located at the point of intersection would therefore occur at both $t' = 1$ s *and* $t' = 0$, which is absurd. Therefore a line connecting events having the same t' coordinate *must* be parallel to another such line.

So if the line connecting all events occurring at the same *time* in the Other Frame is parallel to the diagram x' axis and the line connecting all events occurring at the same *place* in that frame is parallel to the t' axis, we find the coordinates of an event in the Other Frame by drawing lines through the event that are *parallel* to the t' and diagram x' axes (and *not* perpendicular to them). The places where these lines of constant x' and t' cross the coordinate axes indicate the coordinates of the event in the Other Frame (see Fig. 6.10)

Finding the coordinate values by dropping *parallels* instead of perpendiculars may seem strange to you, and will probably take some getting used to. Nonetheless, I hope you see from the argument above that dropping "parallels" is the only way to read the coordinates that makes any sense in this case.

6.7 SUMMARY: DRAWING A TWO-OBSERVER SPACETIME DIAGRAM

FIGURE 6.11

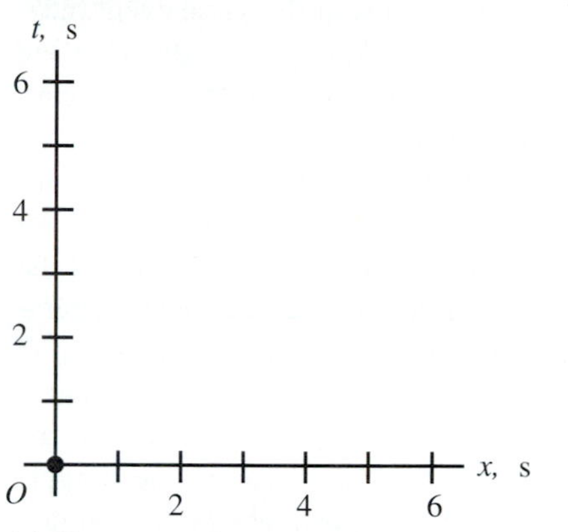

(a) Choose one frame to be the Home Frame. Draw its axes in the usual manner, and indicate the position of the origin event O. Calibrate the axes with some convenient scale.

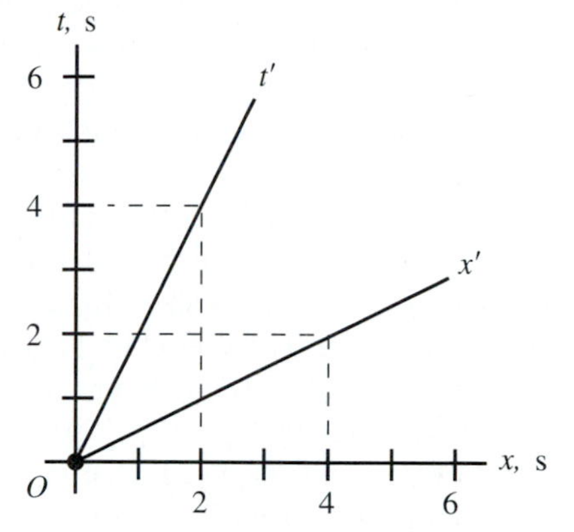

(b) Draw the t' axis of the Other Frame from the origin event O with a slope $1/\beta$ (where β is the x velocity of the Other Frame with respect to the Home Frame). Draw the diagram x' axis of the Other Frame with slope β. This part and subsequent parts are drawn assuming that $\beta = 1/2$.

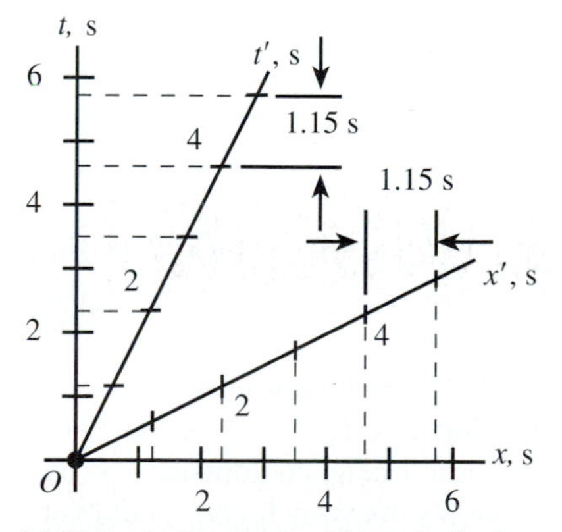

(c) Calibrate the t' axis with marks that are vertically separated by $\Delta t = \gamma \Delta t'$, where $\Delta t' = 1$ s in this case. Calibrate the diagram x' axis with marks having the same spacing. When $\beta = 1/2$, $\gamma \approx 1.15$.

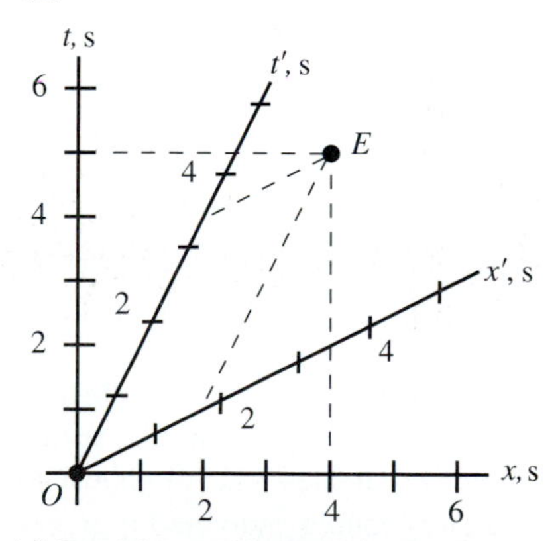

(d) Read the coordinates (in either frame) of any event E by dropping parallels from the event to the appropriate axes. (In this case, E has the coordinates $t \approx 5$ s, $x \approx 4$ s in the Home Frame and $t' \approx 3.4$ s, $x' = 1.7$ s in the Other Frame.)

6.8 THE LORENTZ TRANSFORMATION EQUATIONS

The two-observer spacetime diagram discussed in the preceding sections provides a very visual and intuitive tool for linking the coordinates of an event measured in one inertial reference frame with the coordinates of the same event measured in another inertial frame. Because it is visual in nature, it is much more immediate and less abstract than working with equations. But this tool does lack one thing: the quantitative precision that only equations can provide.

The purpose of this section is to develop a set of *equations* that link the coordinates of an event measured in the Home Frame with the coordinates of the same event measured in the Other Frame. These equations will do *mathematically* exactly what the two-observer diagram does *visually.* These two tools together will enable us to discuss problems in relativity theory with both clarity and precision.

Now, consider an arbitrary event Q, as illustrated in Fig. 6.12. Imagine that we know the coordinates t'_Q and x'_Q of this event in the Other Frame. This means that we can locate an event P which occurs at $t' = 0$ (that is, on the diagram x' axis) and at the same *place* as Q in the Other Frame (that is, $x'_Q = x'_P$). Let the time coordinate separation between events P and Q be Δt_{PQ} and the spatial coordinate separation between O and P be Δx_{OP} in the Home Frame. Proper calibration of the Other Frame axes requires that $\Delta t_{PQ} = \gamma t'_Q$ and $\Delta x_{OP} = \gamma x'_Q$. Also, since the line connecting events P and Q is parallel to the t' axis, its slope must be $1/\beta$, implying that the bottom leg of the triangle involving points P and Q has to have length $\beta\,\Delta t_{PQ}$. Similarly, the slope of the diagram x' axis is β, so the vertical leg of the triangle involving points O and P must have length $\beta\,\Delta x_{OP}$. All these things are shown in Fig. 6.12.

Now, you can see from the diagram that

$$t_Q = \Delta t_{PQ} + \beta\,\Delta x_{OP} = \gamma t'_Q + \gamma\beta x'_Q \qquad (6.9a)$$

$$x_Q = \Delta x_{OP} + \beta\,\Delta t_{PQ} = \gamma x'_Q + \gamma\beta t'_Q \qquad (6.9b)$$

Since the event Q is purely arbitrary, we can drop the subscript and simply say that the Home Frame coordinates t, x of any event can be expressed in terms of the Other Frame coordinates t' and x' of the *same* event as follows:

$$t = \gamma(t' + \beta x') \qquad (6.10a)$$

$$x = \gamma(\beta t' + x') \qquad (6.10b)$$

These equations are called the **inverse Lorentz transformation equations.**

The "just plain" **Lorentz transformation equations,** which express the Other Frame coordinates t', x' of an event in terms of the Home Frame coordinates t, x, can be easily found by solving Eq. (6.10) for t' and x'. The results (which you should verify) are

$$t' = \gamma(t - \beta x) \qquad (6.11a)$$

$$x' = \gamma(-\beta t + x) \qquad (6.11b)$$

These equations can be easily generalized to handle events having nonzero coordinates y and z. We learned in Sec. 4.3 that if two inertial reference frames are in relative motion along a given line, any displacement measured perpendicular to that line has the same value in both frames. Since frames in standard orientation move relative to each other along their common x axis, this means that

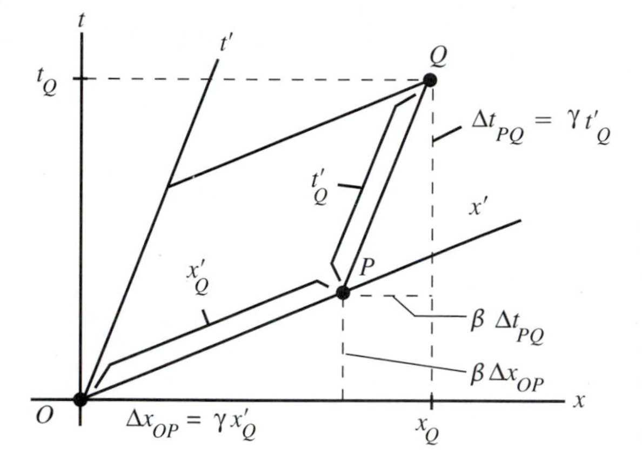

FIGURE 6.12
Pick an arbitrary event *Q*. Then choose event *P* to occur at $t' = 0$ (that is, on the x' axis) and at the same place as *Q* in the Other Frame. Note that since the line connecting events *P* and *Q* is parallel to the t' axis, its slope must be equal to $1/\beta$. Note also that the x' axis has slope β.

$$y' = y \tag{6.11c}$$

$$z' = z \tag{6.11d}$$

Eqs. 6.11 are the relativistic generalization of the galilean transformation equations (1.2).

6.9 COORDINATE TRANSFORMATIONS: VISUAL AND ANALYTICAL

I hope you can see that the derivation in Sec. 6.8 means that the inverse Lorentz transformation equations (or the "plain" Lorentz transformation equations) simply quantify more precisely what you could read from a two-observer spacetime diagram. I chose this method of deriving the Lorentz transformation equations deliberately to drive home the point that the equations simply express *mathematically* what a two-observer spacetime diagram expresses *visually*.

A simple problem will illustrate how these two different methods yield the same results. Consider an event that is observed in the Home Frame to occur at $t = 4.2$ s and $x = 7.0$ s. When and where does this event occur according to observers in an Other Frame that is moving with speed $\beta = 2/5$ with respect to the Home Frame?

Figure 6.13 displays the solution using a two-observer spacetime diagram. Home Frame axes were constructed and scaled, and the event (here marked with an *E*) was plotted with respect to the Home Frame axes according to the coordinates given. Then the Other Frame's t' axis was drawn with slope 5/2, and the diagram x' axis was drawn with slope 2/5. The Other Frame's t' and diagram x' axes were calibrated with marks separated vertically and horizontally (respectively) by $\gamma(1 \text{ s}) = 1.09$ s. Finally, parallels were dropped from event *E* to the t' axis and the diagram x' axis, and the coordinates of *E* were read from the intersection of these parallels with the axes.

Now let us calculate these coordinates using the Lorentz transformation equations found in the last section. With $\beta = 2/5$, $\gamma = 1/\sqrt{1 - \beta^2} = 1/\sqrt{1 - 4/25} \approx 1.09$, so

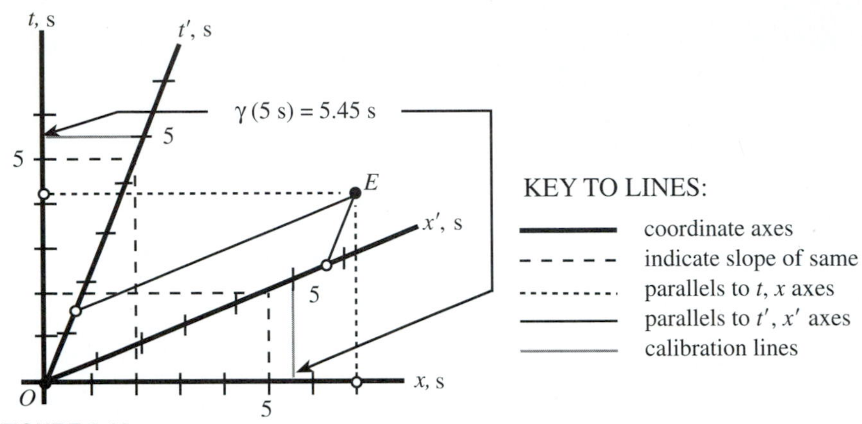

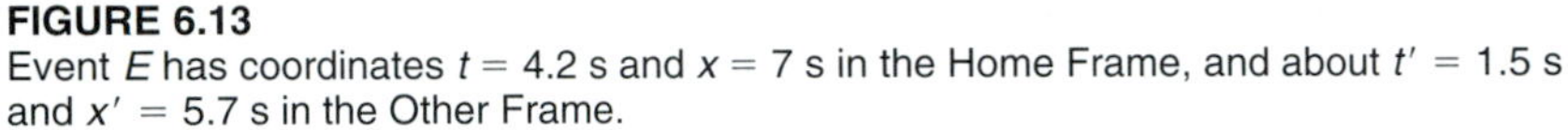

FIGURE 6.13
Event E has coordinates $t = 4.2$ s and $x = 7$ s in the Home Frame, and about $t' = 1.5$ s and $x' = 5.7$ s in the Other Frame.

$$t' = \gamma(t - \beta x) \approx 1.09\,[4.2\text{ s} - (2/5)7.0\text{ s}] \approx 1.5\text{ s} \tag{6.12a}$$
$$x' = \gamma(-\beta t + x) \approx 1.09\,[-(2/5)4.2\text{ s} + 7.0\text{ s}] \approx 5.8\text{ s} \tag{6.12b}$$

which are in substantial agreement with the results above (one should expect an uncertainty of about 3 to 5 percent in results read from even the most carefully constructed spacetime diagram).

6.10 THE TRANSFORMATION OF COORDINATE DIFFERENCES

Often we are not so much interested in the raw coordinates of an event as we are in the coordinate *differences* between two events. Consider a pair of events A and B that are separated by coordinate differences $\Delta t \equiv t_B - t_A$ and $\Delta x \equiv x_B - x_A$ as measured in the Home Frame. What are the corresponding differences $\Delta t' \equiv t'_B - t'_A$ and $\Delta x' \equiv x'_B - x'_A$ as measured in the Other Frame? Applying Eq. (6.11*a*) to t'_A and t'_B separately, we get

$$\begin{aligned}\Delta t' &\equiv t'_B - t'_A = \gamma(t_B - \beta x_B) - \gamma(t_A - \beta x_A)\\ &= \gamma(t_B - \beta x_B - t_A + \beta x_A) = \gamma[(t_B - t_A) - \beta(x_B - x_A)]\end{aligned}$$

Thus

$$\Delta t' = \gamma(\Delta t - \beta\,\Delta x) \tag{6.13a}$$

Similarly,

$$\begin{aligned}\Delta x' &\equiv x'_B - x'_A = \gamma(-\beta t_B + x_B) - \gamma(-\beta t_A + x_A)\\ &= \gamma(-\beta t_B + x_B + \beta t_A - x_A) = \gamma[-\beta(t_B - t_A) + (x_B - x_A)]\end{aligned}$$

Thus

$$\Delta x' = \gamma(-\beta\,\Delta t + \Delta x) \tag{6.13b}$$

$$\Delta y' = y'_B - y'_A = y_B - y_A = \Delta y \tag{6.13c}$$

$$\Delta z' = z'_B - z'_A = z_B - z_A = \Delta z \tag{6.13d}$$

These are the Lorentz transformation equations for coordinate differences. Note that they have the same form as the ordinary Lorentz transformation equations: one simply replaces the coordinate quantities with the corresponding coordinate differences.

The inverse Lorentz transformation equations for coordinate differences are analogous:

$$\Delta t = \gamma(\Delta t' + \beta\,\Delta x') \tag{6.14a}$$

$$\Delta x = \gamma(\beta\,\Delta t' + \Delta x') \tag{6.14b}$$

$$\Delta y = \Delta y' \tag{6.14c}$$

$$\Delta z = \Delta z' \tag{6.14d}$$

6.11 SUMMARY

In this chapter, we have developed two equivalent tools for computing the coordinates of an event in one inertial frame knowing the coordinates in another. The two-observer spacetime diagram illustrates this coordinate transformation *visually.* The construction and interpretation of two-observer diagrams is summarized in Sec. 6.7. The Lorentz transformation equations,

$$t' = \gamma(t - \beta x) \tag{6.11a}$$

$$x' = \gamma(-\beta t + x) \tag{6.11b}$$

$$y' = y \tag{6.11c}$$

$$z' = z \tag{6.11d}$$

express the same coordinate conversion process *mathematically.* The diagrams are more visual and easy to interpret, while the equations are more precise. We will see that these tools complement each other in many applications.

The Lorentz transformation equations and the two-observer spacetime diagrams presented in this chapter complete the set of tools that we have been developing to work with space and time coordinates in the theory of special relativity. Given two events in spacetime, we can now compute the spacetime interval between them, the proper time between them along any path, and the coordinate differences between them in any inertial reference frame, given only the coordinates of the events in any one inertial frame. These tools give us what we need to explore some fascinating implications of special relativity in the next few chapters.

PROBLEMS

6.1 THE DIRECT LORENTZ TRANSFORMATION EQUATIONS. Prove that Eqs. (6.11*a*) and (6.11*b*) follow from Eqs. (6.10*a*) and (6.10*b*).

6.2 DIRECT VS. INVERSE TRANSFORMATION EQUATIONS. Notice that if one changes β to $-\beta$ and exchanges Home Frame for Other Frame coordinates in Eqs. (6.10*a*) and (6.10*b*), one gets (6.11*a*) and (6.11*b*), and vice versa. Write a short paragraph arguing why this in fact makes a great deal of sense. (*Hint:* What is the only difference between the Home Frame and the Other Frame according to our conventions?)

6.3 SPACETIME DIAGRAMS AND THE LORENTZ TRANSFORMATION. An event occurs at $t = 6.0$ s and $x = 4.0$ s in the Home Frame. When and where does this event occur in an Other Frame moving with speed $\beta = 0.50$ in the $+x$ direction with respect to the Home Frame? Answer this question using a two-observer spacetime diagram, and check your work using the appropriate Lorentz transformation equations.

6.4 SPACETIME DIAGRAMS AND THE LORENTZ TRANSFORMATION. An event occurs at $t = 1.5$ s and $x = 5.0$ s in the Home Frame. When and where does this event occur in an Other Frame moving with speed $\beta = 0.60 = 3/5$ in the $+x$ direction with respect to the Home Frame? Does this event occur before or after the origin event in the Other Frame? Answer these questions using a two-observer spacetime diagram, and check your work using the appropriate Lorentz transformation equations. (*Hint:* t' *should come out to be negative.*)

6.5 SPACETIME DIAGRAMS AND THE INVERSE LORENTZ TRANSFORMATION. An Other Frame moves with speed $\beta = 0.60 = 3/5$ in the $+x$ direction with respect to the Home Frame. In the Other Frame, an event is measured to occur at time $t' = 3.0$ s and position $x' = 1.0$ s. When and where does this event occur as measured in the Home Frame? Answer this question using a two-observer spacetime diagram, and check your work using the appropriate inverse Lorentz transformation equations.

6.6 SPACETIME DIAGRAMS AND THE INVERSE LORENTZ TRANSFORMATION. An Other Frame moves with speed $\beta = 0.40 = 2/5$ in the $+x$ direction with respect to the Home Frame. In the Other Frame, an event is measured to occur at time $t' = -1.5$ s and $x' = 5.0$ s. When and where does this event occur as measured in the Home Frame? Does this event occur before or after the origin event in the Home Frame? Answer these questions using a two-observer spacetime diagram, and check your work using the inverse Lorentz transformation equations.

6.7 RELATIVELY CRAFTY. The Federation space cruiser *Execrable* is floating in Federation territory at rest relative to the border of Klingon space, which is 6.0 min away in the $+x$ direction. Suddenly, a Klingon warship flies past the cruiser in the direction of the border at a speed $\beta = 3/5$. Call this event A and let it define time zero in both the Klingon and cruiser reference frames. At $t_B = 5.0$ min according to cruiser clocks, the Klingons emit a parting phasor blast (event B) that travels at the speed of light back to the cruiser. The phasor blast hits the cruiser and disables it (event C), and a bit later (according to cruiser radar measurements) the Klingons cross the border into Klingon territory (event D).

a. Draw a two-observer spacetime diagram of the situation, taking the cruiser to define the Home Frame and the Klingon warship to define the Other Frame. Draw and label the worldlines of the cruiser, the Klingon territory boundary, the Klingon warship, and the phasor blast. Draw and label events A, B, C, and D as points on your diagram.

b. When does the phasor blast hit, and when do the Klingons pass into their own territory, according to clocks in the cruiser's frame? Answer by reading the times of these events directly from the diagram.

c. The Klingon-Federation Treaty states that it is illegal for a Klingon ship in Federation territory to damage Federation property. When the case comes up in Interstellar court, the Klingons claim that they are within the letter of the law, since according to measurements made in their reference frame, the damage to the *Execrable* occurred *after* they had crossed back into Klingon territory: hence they were *not* in Federation territory at the time. Did event C (phasor blast hits the *Execrable*) *really* happen after event D (Klingons cross into Klingon territory) in the Klingon's frame? Answer this question using your two-observer diagram, and check your work with the Lorentz transformation equations.

6.8 TRAINING. Fred sits 65 ns west of the east end of a 100-ns-long train station at rest on the earth. Sally operates a reference frame in a train racing east across the countryside at a speed $\beta = 0.50$. At a certain time (call it $t' = 0$) Sally passes Fred. At that same instant, Fred flashes a strobe lamp (call this event F), which sends bursts of light both east and west. Alan, who is standing at the west end of the station, receives the west-going part of the flash (call this event A), and a bit later (according to clocks in the station) Ellen, who is standing at the east end of the station, receives the east-going flash (call this event E).

a. When do events A and E occur in the station frame? Who sees the flash first (according to clocks in the station), Alan or Ellen?

b. Draw a two-observer spacetime diagram of the situation, showing and labeling the worldlines of Sally, Fred, Alan, Ellen, and the two light flashes. Locate and label events F, A, and E as points on the diagram. Carefully draw and calibrate the t' and x' axes for Sally's train frame.

c. When and where do events A and E occur in Sally's frame? Sally claims that Ellen sees the flash first in her frame. Is this true? Verify your assertions with calculations based on the Lorentz transformation equations.

6.9 TWO-OBSERVER DIAGRAMS FROM BOTH SIDES NOW. (Challenging!) When we have two inertial frames in standard orientation, our convention in preparing a two-observer spacetime diagram is to choose the frame whose relative velocity is in the $+x$ direction to be the Other Frame in the diagram. What if we do it the other way around, i.e., choose the Other Frame to be moving in the $-x$ direction with respect to the Home Frame? Go through the arguments presented in Secs. 6.3 through 6.6 with this change in mind, and then construct a complete and calibrated two-observer spacetime diagram for the case where the Other Frame moves with a speed of $\beta = 2/5$ in the $-x$ direction with respect to the Home Frame. Write a short paragraph describing why you chose to draw and calibrate the diagram as you did.

6.10 DERIVING THE LORENTZ TRANSFORMATION EQUATIONS ANOTHER WAY.[1] We can derive the coordinate difference versions of the Lorentz Transformations in the following way (which is similar to Einstein's own derivation). For the

[1]Adapted from an exercise in E. F. Taylor and J. A. Wheeler, *Spacetime Physics,* San Francisco: Freeman, 1966, p. 68.

sake of argument, we will consider events that occur only along the spatial x axis. First *assume* that the equations have to be linear; i.e., they have the form

$$\Delta t' = A\,\Delta t + B\,\Delta x \tag{6.15a}$$
$$\Delta x' = C\,\Delta t + D\,\Delta x \tag{6.15b}$$

where A, B, C, and D are unknown constants that do not depend on the coordinates at all but only on the relative orientation and velocity of the Home and Other reference frames. (It can be shown that only linear equations like these transform a constant-velocity worldline in the Home Frame into another constant-velocity worldline in the Other Frame. Since a free particle must be measured to move along a constant-velocity worldline in all inertial frames, this is required of any reasonable transformation equation linking two inertial frames.) Then consider the following pairs of events.

a. Consider events E and F that both occur at the spatial origin of the Other Frame so that $\Delta x'_{EF} = 0$. If the Home Frame and Other Frame are in standard orientation, the spatial origin of the Other Frame moves with speed β in the $+x$ direction with respect to the Home Frame, meaning that $\Delta x_{EF}/\Delta t_{EF} = \beta$. Use this to prove that the unknown constants C and D are related as follows: $C = -\beta D$.

b. Imagine a light flash traveling in the $+x$ direction that is emitted at event G and absorbed at event H. The velocity of this light flash has to be observed to be $+1$ in both frames: $\Delta x_{GH}/\Delta t_{GH} = \Delta x'_{GH}/\Delta t'_{GH} = +1$. Show that this implies $C + D = A + B$.

c. Imagine a different light flash emitted at event J and absorbed at event K that travels in the $-x$ direction. The velocity of this light flash has to be observed to be -1 in both frames: $\Delta x_{JK}/\Delta t_{JK} = \Delta x'_{JK}/\Delta t'_{JK} = -1$. Show that this implies $C - D = -A + B$.

d. The three equations from parts *a*, *b*, and *c* above ($C = \beta D$, $C + D = A + B$, and $C - D = -A + B$, respectively), taken together, allow one to express three of the unknowns A, B, C, and D in terms of the fourth. Use these equations to find B, C, and D in terms of A.

e. Finally, the spacetime interval Δs between two events calculated in *either* reference frame must have the same numerical value: we must have $\Delta t^2 - \Delta x^2 = \Delta t'^2 - \Delta x'^2$ (remember that we are only considering events along the x axis here). Use this condition to fix the value of A, and thus all the other constants. You should find that when you plug your results back into Eqs. (6.15), you get the Lorentz transformation equations.[1]

[1]There are many, many ways to derive the Lorentz transformation equations. One of the briefest is described by Alan Macdonald in the May 1981 issue of *Am. J. Phys.* (**49**(5):493). Macdonald's derivation is especially nice because it does *not* require that you assume that the equations are linear. If you draw a spacetime diagram of the situation as you read his article, it will help.

7

LORENTZ CONTRACTION

Physics is essentially an intuitive and concrete science. Mathematics is only a means for expressing the laws that govern phenomena.

Albert Einstein[1]

7.1 OVERVIEW

Two-observer spacetime diagrams may be used to pictorially represent and solve (at least qualitatively) many problems and puzzles involving the theory of special relativity. Because these spacetime diagrams express the relationships between inertial reference frames in a concise and visual manner, solutions using such diagrams often prove clearer and more compelling than mathematical solutions. Even if the precision of a mathematical solution is required, a well-drawn spacetime diagram can speed one on the pursuit of an answer and qualitatively check the result.

In this chapter and the next, we will use two-observer spacetime diagrams (and their mathematical equivalent, the Lorentz transformation equations) to explore the meaning of some of the most peculiar aspects of special relativity. This chapter presents an extended discussion of the phenomenon of Lorentz contraction, including a detailed exploration of the kinds of apparent paradoxes that arise when one misunderstands the true nature of this contraction.

In the next chapter, we discuss the cosmic speed limit (i.e., the fact that nothing can go faster than the speed of light) and derive the Einstein velocity transformation equa-

[1]Quoted (by Maurice Solovine) in A. P. French (ed.), *Einstein: A Centenary Volume,* Cambridge, Mass.: Harvard, 1979, p. 9.

tions [the relativistic generalization of the galilean velocity transformation equations given in Eqs. (1.3)]. In both these chapters, two-observer spacetime diagrams will be used to help make these ideas more "intuitive and concrete."

7.2 THE LENGTH OF A MOVING OBJECT

The basic question that will concern us in this chapter can be stated simply as follows: what exactly do we mean by the "length" of a moving object?

As always, we need an *operational definition* of this word if it is to mean anything; that is, we need to describe exactly how the length of an object can be *measured* in a given inertial frame. In the particular inertial frame where the measuring stick is at rest, it is simple to compare the object to a stationary ruler. But the determination of the length of an object in a frame in which it is observed to be moving presents difficulties that need to be handled carefully.

A reference frame is defined to be an apparatus that measures the spacetime coordinates of events. Our first task in this problem (and indeed most problems in relativity theory) is to rephrase the problem in terms of *events*. In a given reference frame, how might we characterize the length of an object in terms of events?

Let us consider a concrete example. Imagine that we are trying to measure the length of a moving train in the reference frame of the ground. A clock lattice at rest on the ground records the passage of the train through it by describing the motion in terms of events. To be specific, imagine that a certain clock in the lattice records that the *back* end of the train passed by at exactly 1:00:00 P.M. (call this event O). Another clock elsewhere in the lattice records that the *front* end of the train passed by at exactly 1:00:00 P.M. (call this event A). Therefore, we can say that at exactly 1:00:00 P.M., the train lies between the location of the clock registering event O and the clock registering event A. It therefore makes sense to define the length of the train to be equal to the distance between those events as measured in the lattice.

With this image in mind, we will *define* the length of an object operationally as follows:

Definition of Length

The *length* of an object in a given inertial frame is defined to be the *distance* between any two *simultaneous events* that occur at the object's ends.

This expresses the definition of length in the language of events, thus enabling us to use the tools that we have been developing to describe the relationships between events to also talk about the *process* of measuring the length of an object.

7.3 LORENTZ CONTRACTION AND THE TWO-OBSERVER DIAGRAM

Now, consider a measuring stick oriented along the spatial x direction and at rest in the Home Frame. How can we represent such an object on a spacetime diagram? To present the full reality of a measuring stick in spacetime, one must plot the worldline of

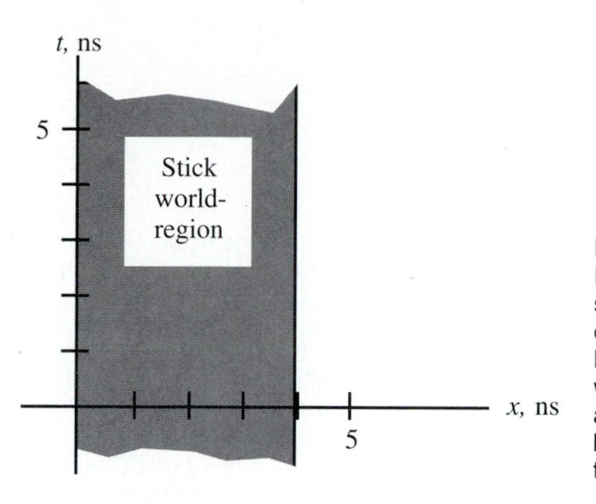

FIGURE 7.1
Part of the world-region of a 4-ns measuring stick that is at rest in the Home Frame with one end at $x = 0$ and the other at $x = 4$ ns. Because it is at rest in the Home Frame, the worldlines of its endpoints are vertical lines, and the worldlines of all of the points in between fill in the shaded region of spacetime shown.

each particle in the stick. Just as a point particle is represented on a spacetime diagram by a curve called a worldline, so a stick is represented by an infinite number of associated worldlines, which one might call a *world-region.* An example of a world-region is shown in Fig. 7.1.

The definition of length given in the previous section yields the expected result when the object in question is at rest. Consider the 4-ns measuring stick of Fig. 7.2, which is at rest in the Home Frame of the diagram. Events O and A lie at the ends of the measuring stick and are simultaneous in the Home Frame: both occur at $t = 0$ in that frame. According to our definition, then, the length of the measuring stick in the Home Frame is the distance between these two events, which, according to Fig. 7.2, is simply 4 ns.

Now consider determining the length of this same measuring stick in the Other Frame, which is moving with speed β in the $+x$ direction with respect to the Home Frame. In that frame, the stick will be measured to move in the $-x$ direction with speed β. Figure 7.3 is a two-observer spacetime diagram showing the Home Frame and Other Frame axes superimposed on the world-region of the measuring stick. (For

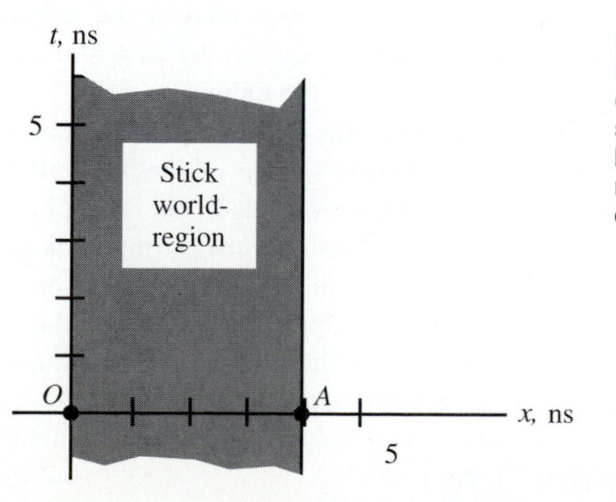

FIGURE 7.2
Events O and A lie along the worldlines of the ends of the measuring stick and occur at the same time in the Home Frame; the distance between these events is the length of the measuring stick in the home frame by definition.

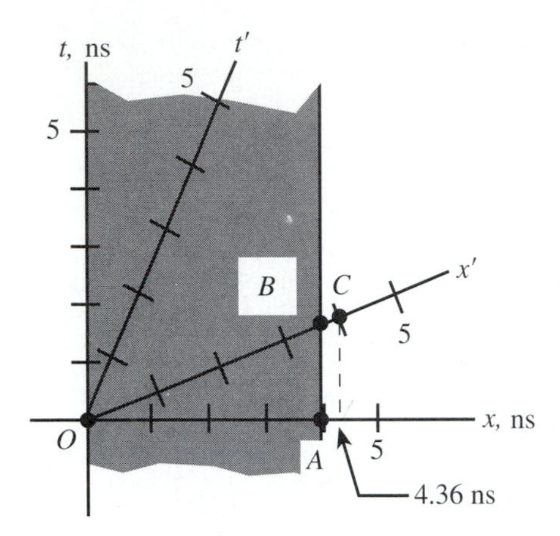

FIGURE 7.3
Events O and B lie at the ends of the measuring stick and occur simultaneously (at $t' = 0$) in the Other Frame. The distance between these events in that frame is thus defined to be the length of the measuring stick in that frame. According to the diagram, this length is about 3.7 ns, not 4 ns. According to the argument in the text, event C, which marks the 4-ns point in the Other Frame, should be farther from the origin event O on the diagram than event B is, implying that $x'_B < 4$ ns.

the sake of being concrete, the diagram is constructed assuming that the relative speed of the frames is $\beta = 2/5$.) In the Other Frame, it is not event A but B (as shown on the diagram) that is simultaneous with O and lies at the other end of the measuring stick (both O and B lie on the diagram x' axis, so both occur at $t' = 0$ in the Other Frame). This means that the length of the measuring stick as observed in the Other Frame is *defined* to be the distance between events O and B as measured in that frame.

But as the calibrated axes of the Other Frame show, the distance between O and B in that frame is *less* than 4 ns! We can see that this *must* be so as follows. Consider the 4-ns mark on the diagram x' axis (the event labeled C) on the spacetime diagram. According to the standard method of calibrating diagram axes, this mark must be separated from the origin event by a horizontal displacement $\Delta x_{OC} = \gamma \Delta x'_{OC} = (1.09)(4 \text{ ns}) \approx$ 4.36 ns. This means that event B must be closer to the origin event than the mark event C at $x' = 4.0$ ns, which in turn implies that $\Delta x'_{OB} < 4$ ns! So the right edge of the measuring stick, which is always exactly 4 ns from the spatial origin of the Home Frame, intersects the diagram x' axis at an event B that is *closer* to the origin event than the 4-ns mark on the diagram x' axis, implying that the stick is measured in the Other Frame to have a length of *less* than 4 ns. In fact, you can see that in the case shown, the stick will be determined to have a length of about 3.7 ns in the Other Frame.

We can also easily check this result with the help of the Lorentz transformation equations. In the Other Frame, the length L of the measuring stick is *defined* to be the distance $\Delta x'$ between two *simultaneous* events occurring at the ends of the stick, i.e., events for which $\Delta t' = 0$. Assuming we know that the length of the measuring stick in the Home Frame is $L_R = 4.0$ ns and that it is at rest in that frame, then the Home Frame distance between *any* pair of events that occur at the opposite ends of the measuring stick must be $\Delta x = L_R = 4.0$ ns (see Fig. 7.3). One of the inverse Lorentz transformation equations for coordinate differences [Eq. (6.14*b*)] says that

$$\Delta x = \gamma(\beta \, \Delta t' + \Delta x') \tag{7.1}$$

In the case at hand, we are looking for $L = \Delta x'$ knowing $\Delta x = L_R$ and $\Delta t' = 0$ for the events in question. Therefore, dropping the $\Delta t'$ term and solving for $\Delta x'$, we get

$$L = \Delta x' = \frac{L_R}{\gamma} = L_R\sqrt{1 - \beta^2} \tag{7.2}$$

Plugging in the relevant numbers in this case, we get

$$L = \sqrt{1 - (2/5)^2}(4.0 \text{ ns}) = 3.7 \text{ ns} \tag{7.3}$$

in agreement with the result displayed by Fig. 7.3.

We see that if we accept the definition of length given in Sec. 7.2 (and how *else* can we define the length of a moving object?), we are confronted with the fact that the length of an object is a *frame-dependent* quantity: its value depends on which inertial reference frame one chooses to make the measurement. Equation (7.2) implies that the length of an object measured in a frame in which the object is moving will always be *smaller* than the value of its length in the frame in which it is at rest. This phenomenon is called **Lorentz contraction.**

Equation (7.2) can be used in general to compute the magnitude of the measured contraction, as long as one remembers that L_R stands for the length of the stick measured in the frame in which it is at rest and L stands for the contracted length of the stick measured in an inertial frame that moves with speed β with respect to the stick's rest frame.

7.4 WHAT CAUSES THE CONTRACTION?

The discussion above shows that this contraction effect has nothing to do with any "magical influence" that causes objects to physically compress when they are moving. The physical reality of the measuring rod (as represented by its world-region on the spacetime diagram) remains the same, no matter what reference frame one uses to describe it. The fundamental reason why observers in different inertial frames will measure the same object to have different lengths is that the observers disagree about clock synchronization and therefore disagree about which events mark out the ends of the object "at the same time" (e.g., observers in the Home Frame use events O and A shown in Fig. 7.3, while observers in the Other Frame use events O and B). We see then that the phenomenon of Lorentz contraction has its origin in the problem of clock synchronization!

Nonetheless, the idea that the same object can be measured to have different lengths in different inertial frames may be hard to accept. Yet we are not at all surprised by the analogous behavior of geometric objects on a two-dimensional plane. Let me illustrate. Imagine that we want to determine the east-west width of a road running in a roughly northerly direction on the surface of the earth. Two different surveyors set up differently oriented coordinate systems and make this measurement. Is it surprising that they get different results (see Fig. 7.4)?

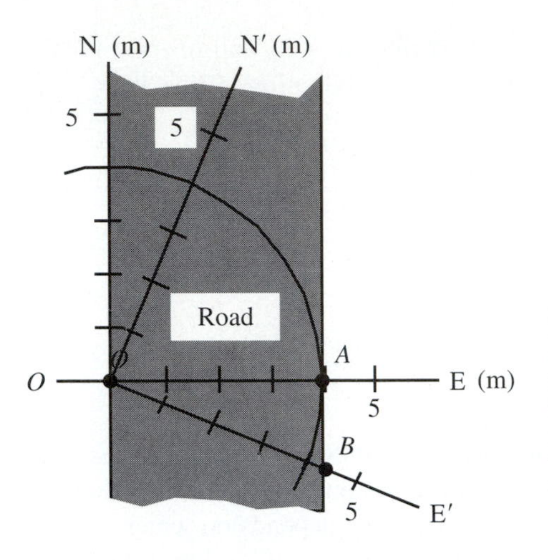

FIGURE 7.4
Points O and A span the east-west width of the road in the Home coordinate system, while points O and B span the east-west width in the Other system. The circle shown connects all points that lie 4 ns from the origin. From the picture it is clear that the road has a greater east-west width in the Other coordinate system than in the Home coordinate system.

The east-west width of the road shown in Fig. 7.4 is seen to be greater in the Other coordinate system than in the Home system. Has the road magically expanded for the surveyor who laid out the Other coordinate system? Hardly! The physical reality of the road does not change just because we change the coordinate system in which we measure it. But because the two surveyors cannot agree on which two points that lie along the sides of the road also lie on an east-west line, they will measure the east-west width of the road to be different.

We do not find this problematic or even unexpected. Now, we might say that if you are going to measure the "true" width of the road, you should measure it using a coordinate system in which the road runs parallel to the y axis so that the x axis is *perpendicular* to the road. In that special coordinate system the width of the road will have its "true" value (which is *shorter* than the value of the same measured in any secondary coordinate system).

Similarly, we might say that to measure the "true" length of an object in spacetime, we should measure its length in the inertial frame in which it is at rest. This "true length" (sometimes called **proper length**) of the object will be *longer* than the value measured in any other inertial frame. For clarity's sake, let us refer to this length as the **rest length** of the object.

7.5 LORENTZ CONTRACTION AND THE PRINCIPLE OF RELATIVITY

But, you might ask, does not this Lorentz contraction effect violate the principle of relativity? We have seen that an object at rest in the Home Frame is measured to have a shorter length in the Other Frame. Does this not imply that there is a physically measurable distinction between the two frames, a distinction that would violate the requirement that all inertial frames be equivalent when it comes to the laws of physics?

The principle of relativity does *not* require that measurements of a specific object or of a set of events have the same values in all reference frames. What the principle *does* require is that if we do exactly the same *physical experiment* in two different inertial reference frames, we will get exactly the same result (otherwise, the laws of physics that predict the outcome of the experiment will be seen to be different in the different frames). Now, we have seen that if we take a measuring stick that is at rest in the Home Frame and 4 ns long in that frame and measure its length in the Other Frame, we will find it to be Lorentz-contracted to 3.7 ns in length. The principle of relativity *does* require that if we perform the *same* experiment in the Home Frame, we should get the *same* result; that is, if we take a 4.0-ns measuring stick at rest in the Other Frame and measure its length in the Home Frame, we should find the stick to be Lorentz-contracted to 3.7 ns.

Figure 7.5 shows that the phenomenon of Lorentz contraction is consistent with the principle of relativity. The worldlines of the ends of a measuring stick that is at rest in the Other Frame will be parallel to the t' axis, as shown. Events O and D mark out the ends of the measuring stick at time $t = 0$ in the Home Frame. The distance between these events (i.e., the length of the measuring stick in that frame) is seen to be about 3.7 ns, as expected.

Again we can also easily check this with the help of the Lorentz transformation equations. In the Home Frame, the length L of the measuring stick is *defined* to be the distance Δx between *simultaneous* events occurring at the ends of the stick, i.e., events for which $\Delta t = 0$. Assuming we know that the length of the measuring stick in the Other Frame is $L_R = 4.0$ ns and that it is at rest in that frame, then the secondary-frame distance between *any* pair of events that occur at the opposite ends of the measuring stick must be $\Delta x' = L_R = 4.0$ ns. One of the Lorentz transformation equations for coordinate differences [Eq. (6.13*b*)] says that

$$\Delta x' = \gamma(-\beta\, \Delta t + \Delta x) \tag{7.4}$$

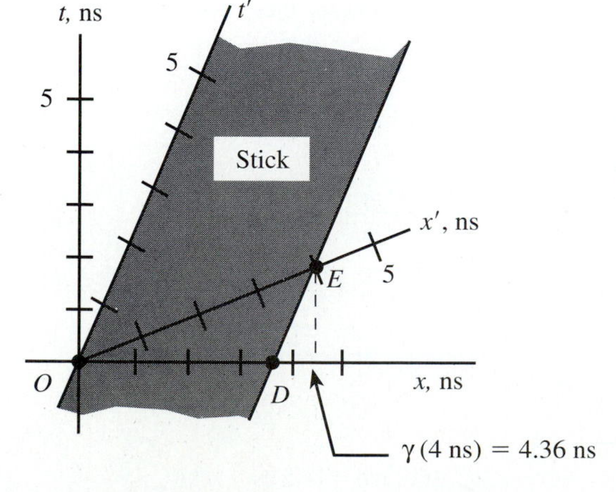

FIGURE 7.5
This measuring stick has a rest length of 4 ns, because events O and E occur at opposite ends of the measuring stick simultaneously in the Other Frame (where the measuring stick is at rest) and are 4 ns apart in that frame. Events O and D lie at the ends of the measuring stick and occur simultaneously in the Home Frame. The distance between these events in that frame is thus defined to be the length of the measuring stick in that frame. According to the diagram, this length is about 3.7 ns, not 4 ns.

In the case at hand, we are looking for $L = \Delta x$ knowing $\Delta x' = L_R$ and $\Delta t = 0$ for the events in question. Therefore, dropping the Δt term and solving for Δx, we get

$$L = \Delta x = \frac{L_R}{\gamma} = L_R\sqrt{1 - \beta^2} = \sqrt{1 - (2/5)^2}(4.0 \text{ ns}) = 3.7 \text{ ns} \tag{7.5}$$

in agreement with the result displayed by Fig. 7.5.

7.6 THE BARN AND POLE PARADOX

The predictions of the theory of relativity are sufficiently counterintuitive that it is easy (as a result of fuzzy thinking) to invent situations based on the ideas of relativity that at first appear to be paradoxical. One of the most famous, the so-called *twin paradox,* was addressed in Chap. 5. In this section, we will examine another famous apparent paradox, generally known as the *barn and pole paradox,* based on the idea of Lorentz contraction.

Consider the following problem. Imagine a pole carried by a pole-vaulter who is running along the ground at a speed $\beta = 3/5$. In the frame of the runner, the pole is at rest (of course): let us assume that it has a rest length of 10 ns. An observer on the ground is moving with speed β with respect to the rest frame of the pole and so will measure the pole to be Lorentz-contracted to a length of only $\sqrt{1 - \beta^2}\,L_R = \sqrt{1 - 9/25}(10 \text{ ns}) = \sqrt{16/25}(10 \text{ ns}) = (4/5)(10 \text{ ns}) \approx 8 \text{ ns}$ (see Fig. 7.6). As the runner presses on, she runs through a barn that also happens to be 8 ns long as measured in the ground frame. Since both the pole and the barn are 8 ns in the ground frame, there is an instant of time in that frame in which *the pole is entirely enclosed by the barn.*

FIGURE 7.6
The world-region of a 10-ns pole carried at a speed of 3/5 relative to the ground frame (the Home Frame in this diagram). Events O and A mark out the ends of the pole at $t = 0$ in the ground frame. The length of the pole in the ground frame is thus the distance between these events, or about 8 ns according to the diagram. For $\beta = 3/5$, $\gamma = 5/4 = 1.25$.

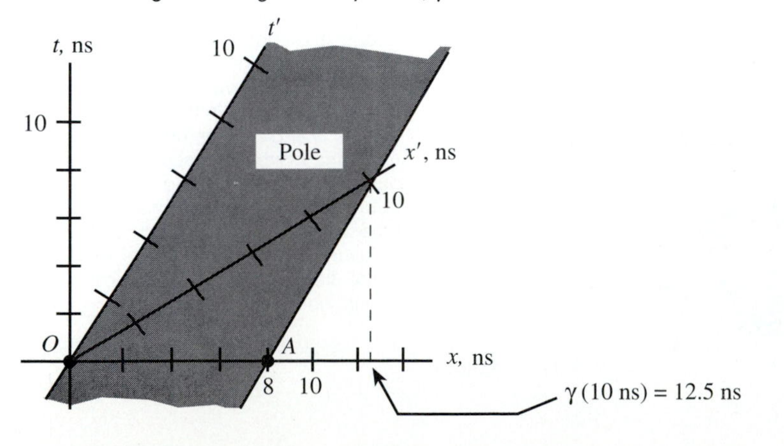

But now look at the situation from the perspective of the runner. In her frame, the pole is at rest and has its normal length of 10 ns. She sees the *barn* to be moving relative to her at a speed of 3/5, and so it is the *barn* that is Lorentz-contracted to 4/5 of its ground-frame length to a length of (4/5)(8 ns) = (32/5) ns = 6.4 ns. Thus the paradox: How can a barn that is 6.4 ns long ever enclose a 10-ns pole?

This apparent paradox results from a naive application of the idea that "moving objects are contracted" without really understanding *why* objects are measured to be contracted and how exactly the length of a moving object is measured. We will see that the apparent paradox is resolved if we carefully consider the precise meaning of the words in the paradox above.

The first step in resolving this problem (and virtually every other problem in special relativity) is to rephrase the problem in terms of *events*. Let us call the arrival of the front end of the pole at the front end of the barn event F. Call the arrival of the back end of the pole at the back of the barn event B. To say that there is an instant at which the barn encloses the pole is to say that events F and B are simultaneous (see Fig. 7.7).

Figure 7.8 shows a two-observer spacetime diagram for this problem. I have chosen event B to be the spacetime origin event (for convenience) and have chosen the ground frame to be the Home Frame of the diagram. I have also taken the ground observer's description of the events to be truthful: events B and F *do* occur simultaneously in the ground frame, and the pole is enclosed by the barn at time $t = 0$ in the ground frame. Notice also that the diagram does bear out the assertion of the runner that the barn is indeed about 6.4 ns long in her frame: events B and C are simultaneous in the runner's frame and lie at the ends of the barn, so the distance between them is thus defined to be the length of the barn in that frame. The diagram shows that this length is indeed about 6.4 ns.

So what is the solution to the paradox? The diagram shows that *the runner never observes the pole to be enclosed by the barn.* Event F is *not* simultaneous with event B in the frame of the runner: rather F is simultaneous with event D (note that the line connecting F and D is parallel to the x' axis). This means that event F (front end of pole reaches front end of barn) occurs about 6 ns *before* event B (back end of pole reaches back end of barn) in the frame of the runner. At the same time as event F occurs in the runner's frame (i.e., at $t' = -6$ ns), you can see from the diagram that the back end of the pole is still sticking way out to the rear of the barn. When event B finally occurs (at $t' = 0$), the front end of the pole is sticking some distance out beyond the front of the barn. (Remember that when we say "at the same time" as an event in

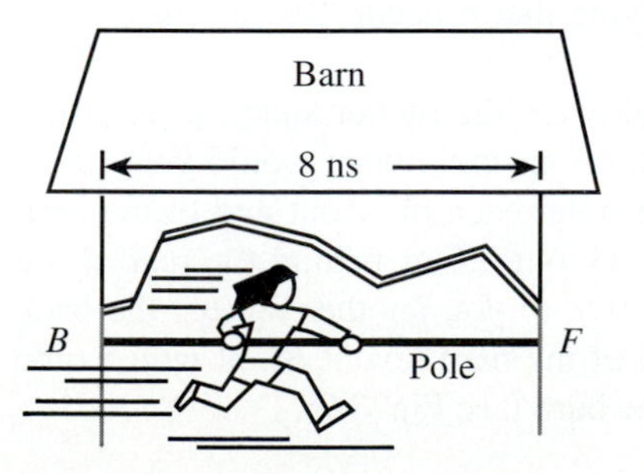

FIGURE 7.7
The pole and barn problem as seen in the reference frame of the ground. Events F (front of pole meets front of barn) and B (back of pole meets back of barn) are simultaneous in that frame.

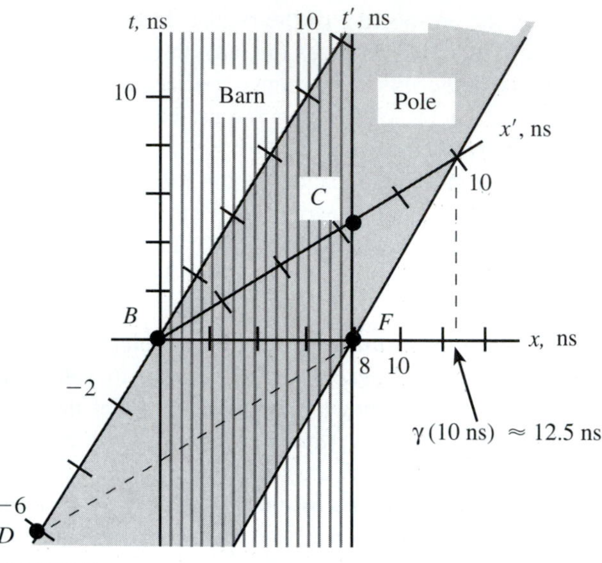

FIGURE 7.8
Solution to the barn and pole paradox. Events B and C mark the ends of the barn at $t' = 0$ in the runner's frame. These events are about 6.4 ns apart, so the barn is indeed about 6.4 ns long as measured in the runner's frame. But note that events B and F are *not* simultaneous in the runner's frame. Indeed, F occurs at the same time as event D, or about 6 ns before B (note that the line connecting events D and F is parallel to the x' axis). When $\beta = 3/5$, $\gamma = 5/4 = 1.25$.

the runner's frame, we are considering events lying on a line parallel to the diagram x' axis.)

We can verify that event F occurs before B in the runner's frame with a quick calculation using the Lorentz transformation equations. In the Home Frame, the coordinate differences between events F and B must be $\Delta t_{BF} = 0$ and $\Delta x_{BF} \equiv x_F - x_B = 8 \text{ ns} - 0 = 8$ ns. The factor $\gamma = 1/\sqrt{1 - \beta^2} = 5/4$ in this case. Therefore, in the runner's frame, we have

$$\Delta t'_{BF} = \gamma(\Delta t_{BF} - \beta\,\Delta x_{BF}) = (5/4)[0 - (3/5)(8 \text{ ns})] = -(3/4)(8 \text{ ns}) \approx -6 \text{ ns} \tag{7.6}$$

Since $\Delta t'_{BF} \equiv t'_F - t'_B$ and $t'_B \equiv 0$, $t'_F = -6$ ns, implying that F occurs about 6 ns of time before B in the runner's frame, as claimed.

Now let us think about this for a minute. If *you* were the runner and you were told that you were about to run a 10-ns pole through a 6.4-ns barn, what would you *expect* to see? First you would see the front end of your pole reach the front end of the barn (event F). At this time, your 10-ns pole would stick out 3.6 ns behind the rear of the 6.4-ns barn. After the barn moves backward relative to you another 3.6 ns, the back end of your pole will coincide with the back end of the barn (event B), at which time the front of the pole sticks out 3.6 ns in *front* of the barn (see Fig. 7.9).

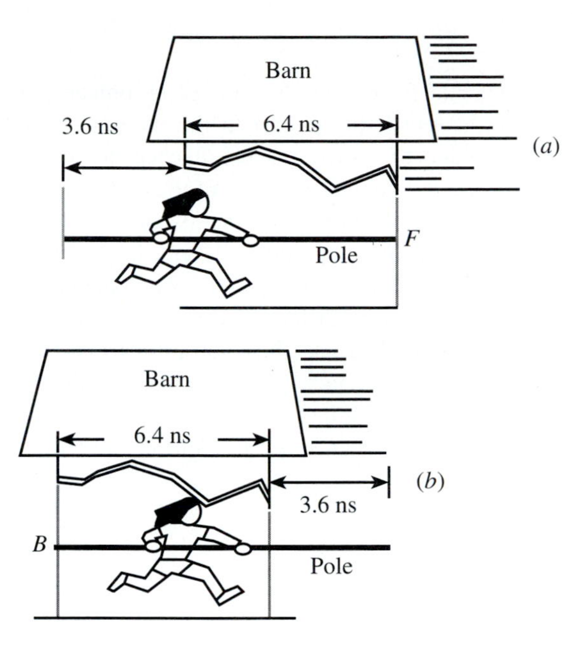

FIGURE 7.9
(*a*) The view from the runner's frame. Event *F* occurs first, at which time the pole sticks 3.6 ns out the rear of the barn. (*b*) Event *B* occurs next, at which time the pole sticks 3.6 ns beyond the front of the barn. The time between events *F* and *B* is the time that it takes the barn to move a distance of 3.6 ns at a speed of $\beta = 3/5$ in the runner's frame.

How long before event *B* should event *F* occur? The time between these events should be the time required for the barn to move backward a distance of 3.6 ns at a speed of $\beta = 3/5$ (in the runner's frame), or

$$\Delta t = \frac{\Delta x}{\beta} = \frac{3.6 \text{ ns}}{(3/5)} = (3.6 \text{ ns})(5/3) = 6 \text{ ns} \tag{7.7}$$

which is the time between the events indicated on spacetime diagram in Fig. 7.8. In fact, you should go over that diagram very carefully and convince yourself that the description given above is indeed exactly what the runner will observe in her reference frame.

The point is that nothing strange or weird happens in either frame. In the barn frame, observed events are consistent with the interpretation that an 8-ns pole is being carried through an 8-ns barn. In the runner's frame, the time relationship between the same events is consistent with the interpretation that a 10-ns pole *really is* being carried through a 6.4-ns barn. We see nothing like "a 10-ns pole enclosed by a 6.4-ns barn." The apparent paradox in the problem as it was stated is based on an unstated and erroneous assumption that if events *F* and *B* were simultaneous (i.e., the pole is enclosed in the barn) in the ground frame, the events will also be simultaneous in the runner's frame. But when we remember that the coordinate time measured between two events will *not* in general be the same in different frames, the paradox dissolves. Excepting the phenomenon of Lorentz contraction itself, *nothing* unusual is seen to happen in either frame.

7.7 SUMMARY

In this chapter, we have discussed the fact that the spatial "length" of an object is a *frame-dependent* quantity, and will be measured to be *longest* in the inertial frame where the object is at rest. This fact is *not* due to some strange influence that compresses objects when they move; it is instead a logical consequence of the definition of the length of a moving object in an inertial reference frame (as "the distance between two simultaneous events occurring at opposite ends of the object") and the fact that events that are simultaneous in one reference frame are not generally simultaneous in another reference frame. Therefore, two observers in different frames measure the same object to have different lengths ultimately because those observers disagree about the synchronization of clocks and thus disagree about what pair of "simultaneous events" should be properly used to measure the length of the object.

In a frame where an object is observed to move with a speed β, its observed length will be

$$L = L_R\sqrt{1-\beta^2} \tag{7.2}$$

This can be shown (approximately) using a two-observer spacetime diagram and (exactly) using the Lorentz transformation equations to link the coordinate separation between two simultaneous events as measured in the original frame with the coordinate differences between those events in the object's own reference frame.

Just as the careless use of the idea that "moving clocks run slow" can lead to unnecessary apparent paradoxes, careless use of the idea that "moving objects contract" without understanding exactly how to describe the measurement of length in terms of events can also lead to apparent paradoxes similar to the barn and pole paradox. These apparent paradoxes can generally be resolved by drawing a carefully constructed two-observer spacetime diagram of the situation and looking carefully in the statement of the paradox for the hidden (incorrect) assumption that events simultaneous in one reference frame are simultaneous in all frames.

PROBLEMS

7.1 LORENTZ CONTRACTION. An observer at rest with respect to the sun measures the diameter of the earth (in the direction of its motion) as it swings by in its orbit. How many centimeters shorter is the earth's diameter in this frame than its rest diameter of 12,760 km? [*Hint:* Use the binomial approximation, Eq. (5.4).]

7.2 LORENTZ CONTRACTION. How fast would an object have to be moving in a given frame if its measured length in that frame is half its rest length?

7.3 LORENTZ CONTRACTION. How fast would an object have to be moving in a given frame if its measured length is to be significantly different from its rest length? (Assume that you can measure the length of the object to 1 part in 10,000, that is, to four significant figures.)

7.4 PARTICLE DECAY PARADOX. Consider the experiment described in Prob. 4.8. In that problem, particles travel 2.08 km between detectors at $v = 0.866$, which takes 8.0 μs as measured by laboratory clocks. Since the half-life of the particles involved is 2.00 μs, we might naively expect only about one-sixteenth of the particles to survive the trip. But it was shown in that problem that the particle clocks only measure 4.0 μs for the trip between the detectors, and thus about one-fourth of the particles actually survive the trip.

But now consider how this all looks to an observer traveling with one of the particles. In the particle's frame, the laboratory and the detectors appear to be moving past at a speed of 0.866. In 4.00 μs (as measured by the particle's clock), the laboratory will only move by a distance of 1.04 km at that speed, so the particles will only see *half* the distance between the detectors go by. But laboratory observers claim that by the time the particles' clock read 4.00 μs, they have covered the *full* distance between the detectors. Is this not a paradox?

Resolve the apparent paradox by considering the phenomenon of Lorentz contraction. How far apart are the detectors in the particle frame? How does this resolve the apparent paradox?

7.5 ANOTHER VIEW ON LENGTH CONTRACTION. We *might* define how to measure the length of an object moving in a given inertial frame as follows: (1) Measure the object's *speed* in the frame, (2) measure the time that it takes the object to pass a given point in the frame, and (3) multiply the speed by the time to get the length. Show that this method yields the same results as the method described in the text using the following procedure:

a. Draw a quantitatively accurate spacetime diagram of a moving spacecraft 100 ns in length that is moving at a speed of $\beta = 3/5$ with respect to the Home Frame. Pick two events A and B that occur at opposite ends of the ship and at the same time in the Home Frame. Show that if the ship's length is defined to be the distance between A and B, it is measured in the Home Frame to be Lorentz-contracted. Estimate the magnitude of the contracted length. (This is the method used in the text.)

b. Use Eq. (7.2) to find the length of the ship as measured in the Home Frame.

c. Now mark two events C and D on your spacetime diagram that represent, respectively, the events of the front end and the rear end of the spacecraft passing the point $x = 0$ in the Home Frame. From your diagram, estimate the time Δt between these events in that frame, and compute $L = \beta\, \Delta t$. Show that the result is equal to the result found in part *a*.

d. Using the Lorentz transformation equations, the fact that C and D are separated by a distance $\Delta x' = L_R = 100$ ns in the spacecraft's frame, and the fact that $\Delta x_{CD} = 0$ by construction, compute the time interval Δt_{CD}. Show that the length computed using $L = \beta\, \Delta t_{CD}$ is equivalent to the length computed using Eq. (7.1).

7.6 LORENTZ CONTRACTION USING THE METRIC EQUATION. An object moving with a speed β passes a clock at rest in your inertial reference frame. Let event F be the front end of the object passing the clock, and let event B be the rear end of the object passing the clock.

a. Argue that in the frame of the object, the coordinate time between these events has to be equal to L_R/β, where L_R is the rest length of the object.

b. What is the *distance* between these events in the object's frame?

c. Let us define the length of the object in *your* reference frame to be the distance that the object travels in the time it takes to pass by your clock, that is, $L = \beta\, \Delta t$, where Δt is the time measured between events F and B by your clock. Use the metric equation and the information in parts *a* and *b* above to arrive at Eq. (7.2). (*Hint:* Your clock is present at both events.)

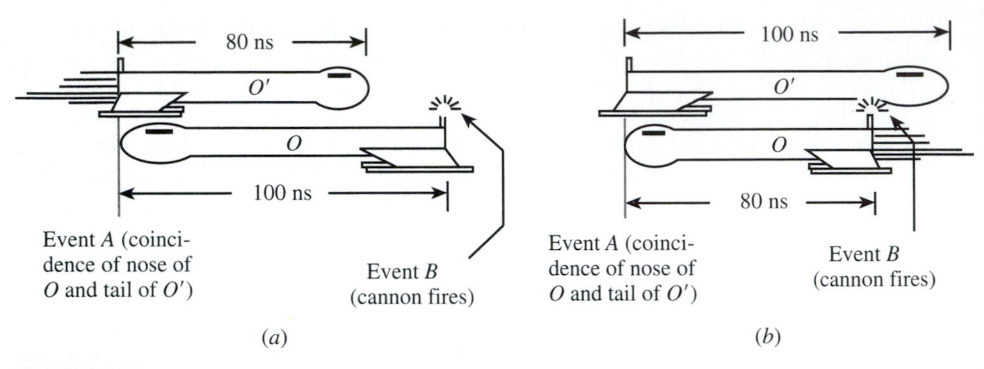

FIGURE 7.10
(a) View from the O frame. *(b)* View from the O' frame.

7.7 SPACE WARS PARADOX.[1] spacecraft of equal rest length $L_R = 100$ ns pass very close to each other as they travel in opposite directions at a relative speed of $\beta = 3/5$. The captain of ship O has a laser cannon at the tail of her ship. She intends to fire the cannon the instant that her bow is lined up with the tail of ship O'. Since ship O' is Lorentz-contracted to 80 ns in O's reference frame, she expects the laser burst to miss the other by 20 ns, as shown in Fig. 7.10*a* (she intends the shot to be "across the bow"). However, to the observer in ship O', it is ship O that is contracted to 80 ns. Therefore, the observer on O' concludes that if the captain of O carries out her intention, the laser burst will strike ship O' 20 ns *behind* the bow, with disastrous consequences (Fig. 7.10*b*). Assume that the captain of O carries out her intentions exactly as described, according to measurements in her own frame, and analyze what *really* happens as follows:

a. Construct a carefully calibrated two-observer spacetime diagram of the situation described. Define event A to be the coincidence of the bow of ship O and the tail of ship O' and event B to be the firing of the laser cannon. Choose A to define the origin event in both frames, and locate B according to the description of O's intentions above. When and where does this event occur as measured in the O' frame, according to the diagram? (You may assume that the ships pass each other so closely that the travel time of the laser burst between the ships is negligible.)

b. Verify the coordinates of B using the Lorentz transformation equations.

c. Write a short paragraph describing whether the cannon burst really hits or not, according to the results you found above. Discuss the hidden assumption in the statement of the apparent paradox, and point out how the drawing in Fig. 7.10 is misleading.

7.8 BULLET-HOLE PARADOX.[2] Two guns are mounted a distance of 40 ns apart on the embankment beside some railroad tracks. The barrels of the guns project outward

[1]Adapted from E. Taylor and J. A. Wheeler, *Spacetime Physics,* San Francisco: Freeman, 1966, pp. 70–71.
[2]Adapted from B. M. Casper and R. J. Noer, *Revolutions in Physics,* New York: Norton, 1972 pp. 363–364.

toward the track so that they almost brush a speeding express train as it passes by. The train moves with a speed of $\beta = 3/5$ with respect to the ground. Suppose the two guns fire simultaneously (as measured in the ground frame), leaving two bullet holes in the train.

a. Let event R be the firing of the rear gun and event F the firing of the front gun. These events occur 40 ns apart and at the same time in the ground frame. Draw a carefully constructed two-observer diagram of the situation, taking the ground frame to be the Home Frame and taking R to be the origin event. Be sure to show and label the axes of the ground and train frames, the worldlines of the guns, the worldlines of the bullet holes that they produce, and the events R and F on your diagram.

b. Using your diagram, argue that the bullet-hole worldlines are about 50 ns apart as measured in the train frame. Verify this by using the Lorentz transformation equations to show that the events R and F occur 50 ns apart in the train frame.

c. In the ground frame, the guns are 40 ns apart. In the train frame, the guns are moving by at a speed of $\beta = 3/5$, and the distance between them will be measured to be Lorentz-contracted to less than 40 ns. Show using the Lorentz contraction formula that the guns are in fact 32 ns apart in the train frame. Describe how this same result can be read from your spacetime diagram.

d. Does this not lead to a contradiction? How can two guns that are 32 ns apart in the train frame fire simultaneously and yet leave bullet holes 50 ns apart in the train frame? Write a paragraph in which you carefully describe the logical flaw in the description of the "contradiction" given in the last sentence. (*Hint:* Focus on the word *simultaneously.*) Describe what *really* is observed to happen in the train frame and thus how it is perfectly natural for guns that are 32 ns apart to make holes 50 ns apart.

7.9 THE CURIOUS CASE OF THE SPACE CADETS. A very long measuring stick is placed in empty space at rest in an inertial frame we will call the Stick Frame. A spaceship of rest length L_R travels along the length of the measuring stick at a speed $\beta = 4/5$ relative to it. Two space cadets P and Q are each equipped with knives and synchronized watches and are stationed at rest on the ship frame in the ends of the spaceship. At a prearranged time, each cadet simultaneously reaches through a porthole and slices through the measuring stick.

a. How long is the spaceship according to the cadets?

b. How long is the spaceship according to observers along the measuring stick (i.e., observers at rest in the Stick Frame)?

c. Use the Lorentz transformation equations to show that observers along the measuring stick would conclude that the cut portion of the measuring stick has length $5L_R/3$.

d. Since the cutting events occur simultaneously in the Spaceship Frame, they do *not* occur simultaneously in the Stick Frame. Use the Lorentz transformation equations to find the time separation of the two cutting events as viewed in the Spaceship Frame.

e. Explain in a short paragraph how it is that two cadets who are only $3L_R/5$ apart (as measured in the Stick Frame) can cut a hunk of measuring stick $5L_R/3$ long if they really cut simultaneously according to their synchronized watches.

7.10 THE TRANSFORMATION OF ANGLES. Consider a meterstick at rest in a given inertial frame (make this the Other Frame) oriented in such a way that it makes an angle of θ' with respect to the x' direction in that frame. In the Home Frame, the Other Frame is observed to move with a velocity of β in the $+x$ direction.

a. Keeping in mind that the distances measured *parallel* to the line of relative motion are observed to be Lorentz-contracted in the Home Frame while distances measured perpendicular to the line of motion are not, show that the angle θ that this meterstick will be observed to make with the x direction in the Home Frame is given by the expression

$$\theta = \tan^{-1} \frac{\tan \theta'}{\sqrt{1-\beta^2}} \tag{7.8}$$

b. What would the length of the meterstick be as measured in the Home Frame?

c. Assume that the meterstick makes an angle of 30° with the x' direction in the Other Frame. How fast would that frame have to be moving with respect to the Home Frame for the meterstick to be observed in the Home Frame to make an angle of 45° with the x direction?

7.11 THE RADAR METHOD. (Challenging!) (Consider an object at rest in the Home Frame. Imagine using the radar method to measure the length of that object in your own frame (the Other Frame). Your frame moves with speed β in the $+x$ direction relative to the Home Frame, so the object appears to be moving with speed β in the $-x$ direction in your frame. At a certain time (event A) you send forth a light flash from a clock in your frame. This flash bounces off a mirror at the far end of the object (event R) and then returns to your clock (event B). If you time this all just right so that the near end of the object passes your clock (event O) at exactly the time halfway between the emission event A and the reception event B (as measured by that clock), then you know that events O and the reflection event R are simultaneous. This means that at that instant, the object lies exactly between the clock and the light flash as it bounces off the mirror. The length L of the object in your frame is thus equal (in SR units) to the time that it takes the light to come back from event R, since light travels at a speed of 1 in all frames.

Now imagine viewing this measurement process from a Home Frame in which your frame is moving with a speed β in the $+x$ direction. The distance between the ends of the object in that frame is L_R. Draw a careful two-observer diagram of the situation as viewed by observers in the Home Frame (let O be the origin event in both frames). Argue that the coordinate distance between events O and A in the Home Frame is $\Delta x = \beta\, \Delta t$, where Δt is the coordinate time measured between those events in the Home Frame. Also argue (using similar triangles on the diagram) that $\Delta t = L_R$. Then use the metric equation to relate the time between events O and A measured in *your* frame (which is equal to L) to the coordinate time Δt measured between the events in the Home Frame, and show that you end up with the same result as that given by the Lorentz contraction equation (7.2).

7.12 LIGHT CLOCKS AND LORENTZ CONTRACTION. (Challenging!) Consider a light clock as shown in Fig. 4.1, except imagine the light clock to be laid on its side so that the light flash travels back and forth parallel to the direction of motion of the clock. Show that the only way this sideways light clock will measure the correct spacetime interval between the events A and B (as any decent clock should) is if the distance between its mirrors is Lorentz-contracted by the amount stated by Eq. (7.2).

8

THE COSMIC SPEED LIMIT

What I'm really interested in is whether God could have made the world in a different way; that is, whether the necessity of logical simplicity leaves any freedom at all.

Albert Einstein[1]

8.1 OVERVIEW

"Nothing can go faster than the speed of light." This fact is a commonly known consequence of the theory of special relativity. But what is it about special relativity that requires this to be true? Are there any loopholes in the argument that might make faster-than-light travel possible?

In Sec. 4.2 and 5.4, we discussed the fact that the metric equation and the proper time equation break down if one attempts to apply them to a clock traveling faster than the speed of light: both equations would predict that the time registered between two events by such a clock would be an imaginary number, which is absurd. This absurdity results from the violation of the $\Delta t^2 > \Delta d^2$ restriction necessary for the derivation of the metric equation. Thus the metric equation really says *nothing* about what a clock traveling at faster than the speed of light will measure.

In fact, the question of what a clock traveling faster than light would measure is a moot issue in special relativity. In this chapter, we will see (with the help of some two-observer diagrams) that it is a consequence of the principle of relativity that not only can no *clock* travel faster than the speed of light, but it is not possible for any object or

[1]Quoted in A. P. French (ed.), *Einstein: A Centenary Volume,* Cambridge, Mass.: Harvard, 1979, p. 128.

even a *message* to travel faster than light! The speed of light thus represents a true "cosmic speed limit."

In this context, we will also have to rethink the galilean velocity transformation. For example, if a particle is traveling at a speed of $0.9c$ in the $+x$ direction in a frame that is itself traveling at a speed of $0.9c$ in the $+x$ direction with respect to our frame, the galilean velocity transformation equations imply that the particle would be measured to have a speed of $1.8c$ in our frame, a clear violation of the cosmic speed limit. The galilean velocity transformation equations are thus clearly wrong (as we already knew from the fact that the speed of light has the same value in every inertial frame). But how can we arrive at the equations that take their place?

In our discussion of these issues in this chapter, we will see that two-observer spacetime diagrams again play a crucial role in vividly illustrating the basic physical issues involved.

8.2 CAUSAL CONNECTIONS AND THE ULTIMATE SPEED

The problem with faster-than-light travel is that it *violates causality.* What do I mean by *causality?* In physics (and more broadly, in daily life), we know that certain events *cause* other events to happen (Fig. 8.1). For example, even couch potatoes know that if you press the appropriate button on the remote control, the TV channel will change. These kinds of causal connections imply that certain causally connected events have an invariant temporal order: for example, we would be deeply disturbed if the TV channel changed just *before* we pressed the remote control button.

FIGURE 8.1
Causal connections.

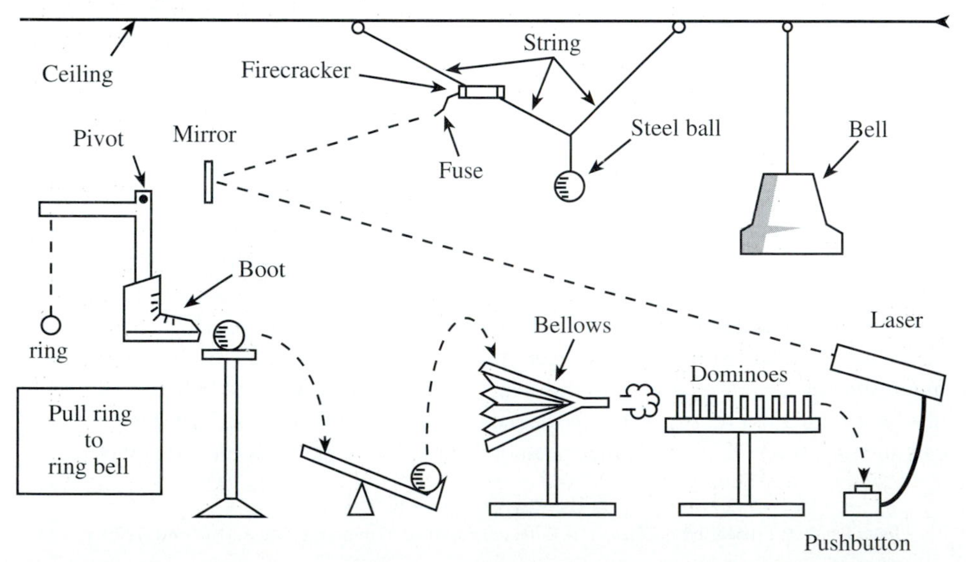

Consider two distinct events (call them P and Q, respectively) such that event P *causes* event Q. That is, Q is *caused* to occur upon reception of some kind of information that P has occurred. This information can be transmitted from P to Q in any number of ways: via some mechanical effect (such as the movement of an object or the propagation of a sound wave), via a light flash, via an electrical signal, via a radio message, etc. Basically, the information can be carried by *any* object or effect that can move from place to place and is detectable.

Let us consider the TV remote control again as a specific example. Imagine that you press a button on your TV remote control handset (event P). The information that the button has been pressed is sent to the TV set in some manner, and in response, the TV set changes channels (event Q). Keep this basic example in mind as we go through the argument that follows.

Now let us pretend that the "causal influence" that connects event P to event Q *can* flow between them at a constant speed v_{ci} *faster* than the speed of light as measured in your inertial reference frame, which we will call the Home Frame (perhaps the TV manufacturer has found some way to convey a signal from the remote to the TV using "Z waves" that travel faster than light). We will show that this leads to a logical absurdity. Choose event P to be the origin event in that frame, and choose the x axis of the frame so that both events P and Q lie along it. (We can always do this: it is just a matter of choosing the origin and orientation of our reference frame. Choosing the frame to be oriented this way is just a matter of convenience.)

Figure 8.2 shows a spacetime diagram (drawn by an observer in the Home Frame) of a pair of events P and Q fitting the description above. Note that if the causal influence flows from P to Q faster than the speed of light, its worldline on the diagram will have a slope $1/v_{ci} < 1$, that is, *less* than the slope of the worldline of a light flash leaving event P at the same time (which is also shown on the diagram for reference).

Now consider Fig. 8.3. In this two-observer spacetime diagram, I have drawn the t' and x' axes for an Other Frame that travels with a speed β with respect to the Home Frame. Note that according to Sec. 6.5 the slope of the diagram x' axis in such a diagram is β. Note also that since the slope of the causal influence worldline is $1/v_{ci} < 1$, it is always possible to find a value of β such that $1/v_{ci} < \beta < 1$, meaning that the slope of

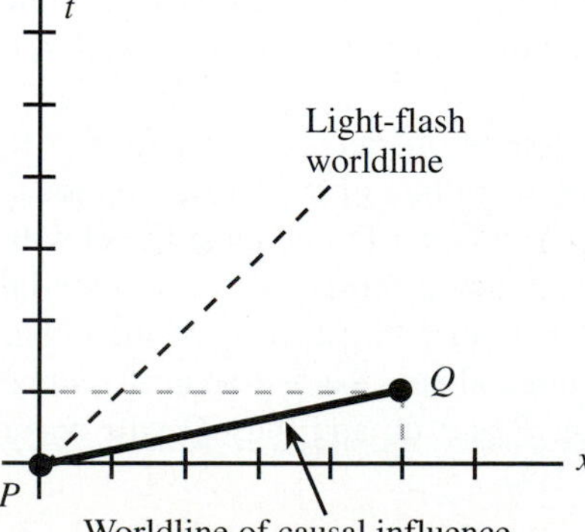

FIGURE 8.2
Imagine that events P and Q are connected by a hypothetical causal influence traveling with a speed v_{ci} faster than the speed of light. In the case shown, P has been chosen (for the sake of convenience) to be the origin event, and the influence travels 5 s of distance for every second of time, so its speed is $v_{ci} = 5$ (that is, five times the speed of light).

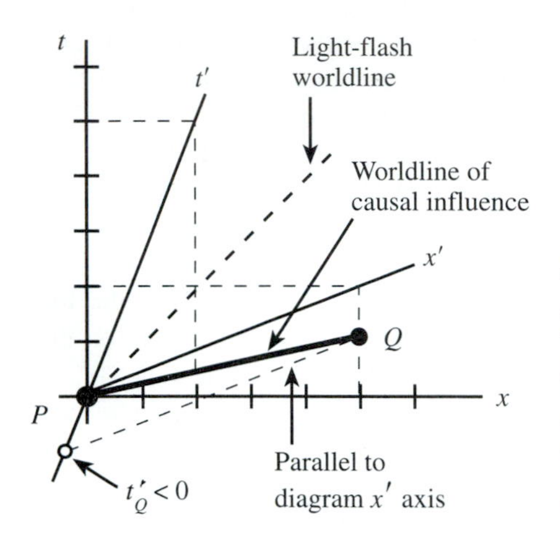

FIGURE 8.3
This two-observer spacetime diagram is the same as in Fig. 8.2 with the addition of the axes for an Other Frame that is moving with a speed β with respect to the Home Frame. In the drawing, β has been chosen to be 2/5, so the slope of the diagram x' axis is $2/5 > 1/5$, which is the slope of the causal influence worldline. Note that in the Other Frame, event Q will be measured to occur *before* event P.

the x' axis lies *between* the light-flash worldline and the causal connection worldline, as shown. In such a frame, event Q will be measured to occur *before* event P, as one can see by reading the time coordinates of these events from the diagram.

Here is the absurdity: in the Other Frame, event P is observed to occur *after* event Q does. This is absurd, because event P is supposed to *cause* event Q. How can an event be measured to occur before its cause? This is not merely a semantic issue, nor is it mere appearance. According to any and every physical measurement that one might make in the Other Frame, event Q will really be observed to occur *before* its "cause" P.

To vividly illustrate the absurdity, consider our TV remote example. If the signal could go from your remote control to the TV faster than light, in certain inertial reference frames, you would observe the TV set to change channels *before* the button was pushed. If this were to happen in your reference frame, you would consider this a violation of the laws of physics (presuming your TV set was not broken). But the laws of physics are supposed to hold in every inertial reference frame. Therefore this observed inversion of cause and effect violates the principle of relativity!

We have only three options at this point. We can reject the principle of relativity and start over at square one. We can radically modify our conception of causality in a way that is yet unknown. Or we can reject the assumption that got us into this trouble in the first place, namely, that a causal influence can flow from P to Q faster than light, i.e., with $v_{ci} > 1$.

The latter option is clearly the least drastic. If information can only flow from P to Q with a speed $v_{ci} \leq 1$ in the Home Frame, then the worldline of the causal influence connecting event P to Q will have a slope $1/v_{ci} > 1$. Any Other Frame must travel with $\beta < 1$, by this hypothesis (since the parts of the reference frame, like any material object, could be the agent of a causal influence). Therefore, the slope β of the Other Frame's diagram x' axis on a spacetime diagram must always be less than the slope $1/v_{ci}$ of the causal connection worldline connecting P and Q, and thus Q will occur after P in *every* Other inertial reference frame (Fig. 8.4).

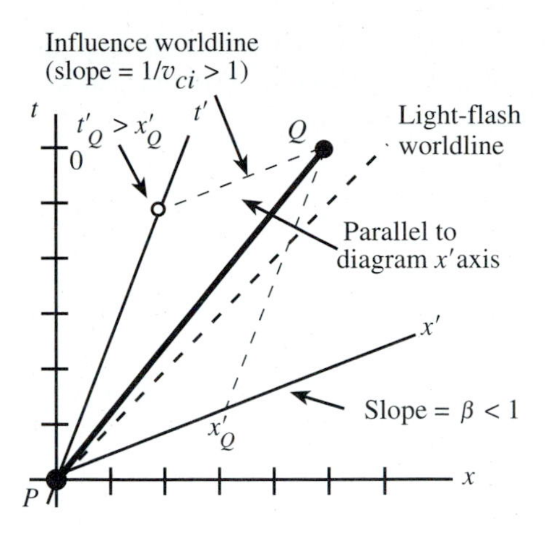

FIGURE 8.4
Assume that the speed v_{ci} of the causal influence between event *P* and *Q* satisfies the restriction $v_{ci} < 1$ in the Home Frame. Consider any Other inertial frame. The slope of the Other Frame's diagram *x*′ axis is equal to the speed of the Other Frame with respect to the Home Frame, which must be less than 1. Thus the slope of line *PQ*, which is $1/v_{ci}$, will always be *greater* than the slope of the *x*′ axis, so *P* will be measured to occur *before Q* in *any* Other inertial frame.

In short, if the speed of reference frames and causal influences is limited to $v_{ci} < 1$, then effects will occur *after* their causes in every inertial reference frame, as required by the principle of relativity when applied to the concept of causality.[1]

In summary:

Theorem: The Cosmic Speed Limit

In order for causality (i.e., the idea that one event can *cause* another event to happen) to be self-consistent, information (i.e., *any* effect representing a causal connection between two events) cannot travel between two events with a speed v_{ci} greater than that of light.

Since anything movable and detectable can carry information (i.e., cause things to happen), this consequence of the principle of relativity applies not only to all physical objects (waves, particles, and macroscopic objects) but indeed to *any* trick or means of conveying a message that might be imagined (e.g., instantaneous changes in a gravitational field, telepathy, magic, whatever.)

So, with a straightforward argument using two-observer spacetime diagrams, we have proved the existence of a cosmic speed limit, an idea having profound physical and and philosophical implications. As usual, this prediction is amply supported by experiment. No particle, object, or signal of any kind has ever been definitely observed to travel at faster than the speed of light in a vacuum. Science fiction fans and space travel buffs who hope for the discovery of faster-than-light travel may hope forever:

[1]At the most basic physical level, the problem with causes following effects is that this would violate the second law of thermodynamics, which requires that the entropy of the universe always increase (or at least remain the same) during a physical process. This law implies that events in certain physical processes will occur in one temporal order but not in the reverse order. Therefore, if the second law of thermodynamics is to be true in all inertial frames, as required by the principle of relativity, then the temporal order of all events that might be linked by that law must be preserved in all inertial frames. "Cause" and "effect" is thus really a colloquial way to talk about the invariant temporal order imposed on events by the second law of thermodynamics.

both the argument (based as it is on the firmly accepted and fundamental ideas of the principle of relativity and the physical reality of causality) and the experimental evidence present a pretty ironclad case for this cosmic speed limit.

8.3 TIMELIKE, LIGHTLIKE, AND SPACELIKE INTERVALS

We are now in a position to understand more fully the true physical nature of the spacetime interval between *any* two events in spacetime. In Sec. 4.2, we saw that for two events whose coordinate differences in a given inertial reference frame are Δt, Δx, Δy, and Δz, the quantity $\Delta s^2 = \Delta t^2 - \Delta x^2 - \Delta y^2 - \Delta z^2$ had a frame-independent value, and that value was equal to the time registered by an inertial clock present at the two events. But to make the proof of the metric equation [Eq. (4.5)] work, it was necessary to assume that $\Delta d \equiv \sqrt{\Delta x^2 + \Delta y^2 + \Delta z^2}$ was *smaller* than Δt so that there was more than sufficient time for a light flash to travel from one event to the other along the length of the light clock. The purpose of this section is to study the meaning of the spacetime interval when $\Delta d > \Delta t$, that is, when this condition is violated.

We have exploited the analogy between spacetime geometry and euclidean plane geometry extensively in the last few chapters. We have noted, though, that the negative signs in the metric equation (which do not appear in the corresponding pythagorean relation) lead to some subtle differences between spacetime geometry and euclidean geometry. Yet another one of these differences is the following. In euclidean geometry the square distance Δd^2 between two points on a plane is necessarily positive:

$$\Delta d^2 = \Delta x^2 + \Delta y^2 \geq 0 \tag{8.1}$$

But taken at face value, the metric equation allows the squared spacetime interval between two events to be positive, zero, or negative, depending on the relative sizes of the coordinate separations Δd and Δt between those events:

$$\Delta s^2 = \Delta t^2 - \Delta d^2$$

Therefore

$$\text{If} \quad \Delta d > \Delta t \qquad \text{then} \qquad \Delta s^2 < 0! \tag{8.2}$$

We see that while there is only one kind of distance between pairs of points on a plane, the possible spacetime intervals between pairs of events in spacetime fall into three distinct categories depending on the sign of Δs^2. These categories are:

If $\Delta s^2 > 0$, the interval between the events is said to be **timelike**.
If $\Delta s^2 = 0$, the interval between the events is said to be **lightlike**.
If $\Delta s^2 < 0$, the interval between the events is said to be **spacelike**.

The reasons for these names will become clear shortly.

The peculiar category here is the spacelike category—there is nothing corresponding to it in ordinary plane geometry (where the squared distance between two events is always positive). What does it mean for two events to have a *spacelike* spacetime interval between them?

First of all, note that events separated by spacelike spacetime intervals certainly do exist. Consider, for example, the case of two events that are measured in a certain inertial frame to occur at the same *time* but at different *locations.* Since the time separation between these events is zero in that reference frame, we have $\Delta s^2 = 0 - \Delta d^2 = -\Delta d^2 < 0$, so the interval between these events is necessarily spacelike. Therefore, the spacelike interval classification is needed if we are to be able to categorize the spacetime interval between arbitrarily chosen events.

As we have already discussed, the squared spacetime interval between two events Δs^2 that appears in the metric equation $\Delta s^2 = \Delta t^2 - \Delta d^2$ has been linked with the frame-independent time measured by an inertial clock present at both events *only* in the case that $\Delta t^2 > \Delta d^2$. For two events for which $\Delta t^2 < \Delta d^2$, it is not clear how one can directly measure the squared spacetime interval between the events at all. For example, for an inertial clock to be present at both events if $\Delta d > \Delta t$, it would have to travel at a speed $v > 1$ in that frame. We have just seen that this is impossible; thus a spacelike spacetime interval cannot be measured by a clock or anything else that travels between the events. Since the proof of the metric equation given in Sec. 4.2 does not handle the case of spacelike intervals, it is not even completely clear that the squared interval $\Delta s^2 \equiv \Delta t^2 - \Delta d^2$ is frame-independent when it is less than zero.

In fact, the squared spacetime interval Δs^2 *does* have a frame-independent value, no matter what its sign is. This can easily be demonstrated by using the Lorentz transformation equations for coordinate differences given by Eq. (6.13). The argument goes like this. Let $\Delta t, \Delta x, \Delta y, \Delta z$ be the coordinate separations of two events measured in the Home Frame, and let $\Delta t', \Delta x', \Delta y', \Delta z'$ be the coordinate separations of the same two events measured in an Other Frame moving with speed β in the $+x$ direction with respect to the Home Frame. Then, Eqs. (6.13) imply that

$$\begin{aligned}
(\Delta t')^2 - (\Delta x')^2 - (\Delta y')^2 - (\Delta z')^2 \\
&= [\gamma(\Delta t - \beta\,\Delta x)]^2 - [\gamma(-\beta\,\Delta t + \Delta x)]^2 - \Delta y^2 - \Delta z^2 \\
&= \gamma^2(\Delta t^2 - \beta\,\Delta x\,\Delta t + \beta^2\,\Delta x^2) - \gamma^2(\beta^2\,\Delta t^2 - \beta\,\Delta x\,\Delta t + \Delta x^2) - \Delta y^2 - \Delta z^2 \\
&= \gamma^2(\Delta t^2 + \beta^2\,\Delta x^2 - \beta^2\,\Delta t^2 - \Delta x^2) - \Delta y^2 - \Delta z^2 \\
&= \gamma^2(1 - \beta^2)(\Delta t^2 - \Delta x^2) - \Delta y^2 - \Delta z^2 \\
&= \Delta t^2 - \Delta x^2 - \Delta y^2 - \Delta z^2
\end{aligned} \tag{8.3}$$

where the last step uses the fact that $\gamma \equiv 1/\sqrt{1 - \beta^2}$. The sign of $\Delta s^2 = \Delta t^2 - \Delta d^2$ is irrelevant to this derivation: thus Δs^2 *always* has the same numerical value in every inertial reference frame, no matter whether it is spacelike, timelike, or lightlike.

How can we measure the value of the spacetime interval between two events separated by a spacelike interval? We cannot use a clock, as we have noted above. In fact, a spacelike spacetime interval is measured with a *ruler,* as we will shortly see.

Let us define the **spacetime separation** $\Delta\sigma$ between two events as follows:

$$\Delta\sigma^2 = \Delta d^2 - \Delta t^2 \tag{8.4}$$

The spacetime separation, so defined, is conveniently real whenever the interval between the events is spacelike. Now note that *if* we can find an inertial reference frame where the events are simultaneous ($\Delta t = 0$), we have

$$\Delta\sigma^2 = \Delta d^2 \Rightarrow \Delta\sigma = \Delta d \qquad \text{(in a frame where } \Delta t = 0) \tag{8.5}$$

Now, I claim that we can *always* find a frame in which $\Delta t = 0$ if the events are separated by a spacelike interval. This can be seen as follows. Imagine that two events occur with coordinate differences Δt and Δd ($> \Delta t$) as measured in the Home Frame. Reorient and reposition the axes of the Home Frame so that the events in question both occur along the spatial x axis, with the later event located in the $+x$-direction relative to the earlier event. (This can be done without loss of generality: we are always free to choose the orientation of our coordinate system to be whatever we find convenient.) Once this is done, $\Delta d = \Delta x$ in the Home Frame.

Now, consider an Other Frame in standard orientation with respect to the Home Frame and traveling in the $+x$ direction with speed β with respect to the Home Frame. According to Eq. (6.13*a*) the time coordinate difference between these events in the Other Frame is

$$\Delta t' = \gamma(\Delta t - \beta\, \Delta x) \tag{8.6}$$

These events will be simultaneous in the Other Frame (that is, $\Delta t'$ will equal zero) if and only if the relative speed of the frames is chosen to be $\beta = \Delta t/\Delta x = \Delta t/\Delta d$. This relative speed β will be less than 1 since $\Delta d > \Delta t$ for our events by hypothesis. In short, given *any* pair of events that are separated by a spacelike interval in some inertial frame (which we are calling the Home Frame), it is possible to find an Other inertial frame moving with speed $\beta < 1$ with respect to the Home Frame in which the two events will be simultaneous (see Fig. 8.5). QED

To summarize, then, if the spacetime interval between two events is *spacelike,* then:

1 It is possible to find an inertial frame in which these events occur at the *same time.*

2 $\Delta\sigma$ is the *distance* between the events in that special frame.

3 Observers in all other inertial reference frames can use Eq. 8.4 to *calculate* $\Delta\sigma$: they will all get the same value for $\Delta\sigma$ that is measured directly in the special frame.

These statements are directly analogous to statements that can be made about events separated by a timelike spacetime interval. If the spacetime interval between two events is *timelike,* then:

1 It is possible to find an inertial frame in which these events occur at the *same place* (this is the frame of the inertial clock that is present at both events).

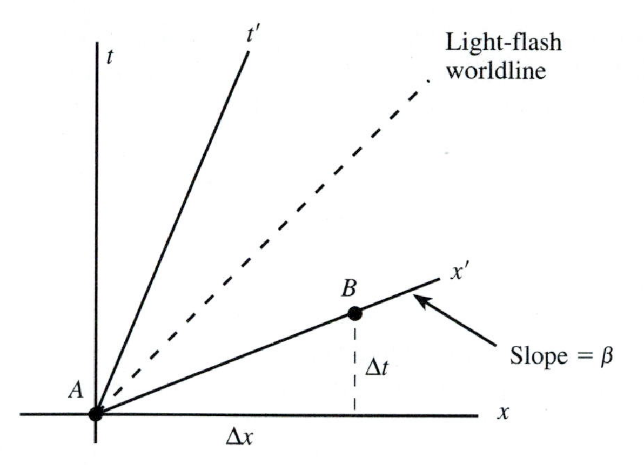

FIGURE 8.5
Given any pair of events A and B separated by a spacelike interval ($\Delta x > \Delta t$) in the Home Frame, it is possible to find an Other Frame in which the events are simultaneous. The speed of this special Other Frame with respect to the Home Frame simply has to have the right value to give the slope of the diagram x' axis the value $\Delta t/\Delta x$, that is, $\beta = \Delta t/\Delta x$. Since $\Delta x > \Delta t$, this $\beta < 1$, so it is always possible to find such a frame.

2 Δs is the *time* between the events in this special frame (i.e., the time measured by that inertial clock).

3 Observers in all other inertial reference frames can use the ordinary metric equation to calculate Δs: they will all arrive at the same value measured directly in the special frame.

Thus there is a fundamental symmetry between spacelike and timelike spacetime intervals, a symmetry that arises because both reflect the same underlying physical truth: It is possible to describe the separation of *any* two events in space and time with a frame-independent quantity Δs^2 (which we will call the **squared spacetime interval**) that is analogous to the *squared distance* between two points in plane geometry. It is simply a peculiarity of the geometry of spacetime that the quantity in spacetime that corresponds to ordinary (unsquared) distance on the plane comes in three distinct flavors (the spacetime *interval* Δs if $\Delta s^2 > 0$ for the events, the spacetime *separation* $\Delta\sigma$ if $\Delta s^2 < 0$, and the lightlike interval $\Delta s = \Delta\sigma = 0$ when $\Delta s^2 = 0$) which are measured in different ways using different tools. But it is important to realize that these three quantities are only different aspects of the same basic frame-independent quantity Δs^2.

We see that timelike intervals are directly measured with a *time*-measuring device (i.e., an inertial *clock* present at both events), while spacelike intervals are directly measured with a *space*-measuring device (i.e., a *ruler* in the special inertial frame where Δt between the events is zero). This is why these interval classifications have the names *timelike* and *spacelike:* the names are intended to tell us whether we have to measure the interval with a clock (because it is *timelike*) or with a ruler (because it is *spacelike*). The lightlike interval classification stands between the other classifications. When the interval between two events is lightlike, we have $\Delta d = \Delta t$, which implies that these events could be connected by a flash of light.

8.4 THE CAUSAL STRUCTURE OF SPACETIME

Now, *because* it is true that the value of the squared spacetime interval Δs^2 is frame-independent no matter what its sign, all inertial observers will agree as to whether the interval between a given pair of events is timelike, lightlike, or spacelike (since if they all agree on the *value* of Δs^2, they will surely all agree on its sign). This means that the spacetime around any event P can be divided up into the distinct regions shown in Fig. 8.6, and every observer will agree about which events in spacetime belong to which region.

Because these regions can be defined in a frame-independent manner, it is plausible that they reflect something absolute and physical about the geometry of spacetime. In fact, *these regions distinguish those events that can be causally connected to P from those that cannot.* Remember that in Sec. 8.2 we found that two events can be causally linked only if $\Delta d \leq \Delta t$ between them: otherwise the causal influence would have to travel between the events faster than the speed of light. This means that every event that can be causally linked with P must have a timelike (or perhaps lightlike) interval with respect to P: such events will lie in the white regions shown in Fig. 8.6.

We can be more specific yet. Since the temporal order of events is preserved in all inertial frames if $\Delta d > \Delta t$ (see Fig. 8.3 and the surrounding text), all events in the *upper* white region in Fig. 8.6 will occur *after P* in every frame (and thus could be caused by P), and all events in the lower white region of Fig. 8.6 will occur *before P* in every frame (and thus could cause P). These regions are referred to as the *future* and *past* of P, respectively.

Events whose spacetime interval with respect to P is spacelike ($\Delta d > \Delta t$) cannot influence P or be influenced by it. We say that these events (which inhabit the shaded region of Fig. 8.6) are causally *unconnected* to event P.

With this in mind, we can relabel the regions in Fig. 8.6 as shown in Fig. 8.7. Remember that every observer agrees on the value of the spacetime interval between event P and any other event, so every observer agrees as to which event belongs in which classification. The structure illustrated is thus an intrinsic, frame-independent characteristic of the geometry of spacetime.

Now, the boundaries of the regions illustrated in Fig. 8.7 are light-flash worldlines. If we consider two spatial dimensions instead of one, an omnidirectional light flash is seen as an ever-expanding ring, like the ring of waves formed by the splash of a stone

FIGURE 8.6
The regions of spacetime relative to an event P.

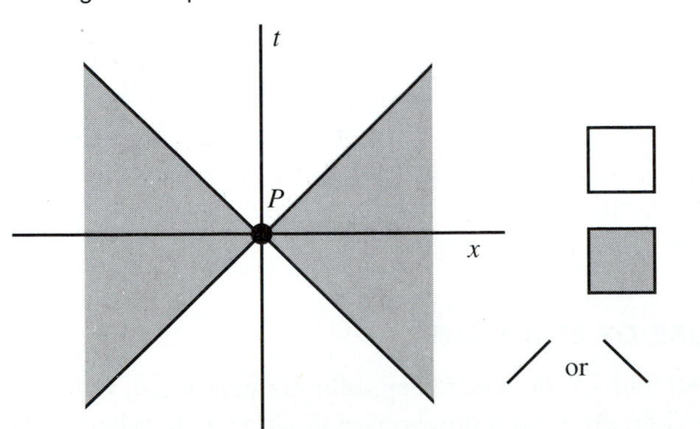

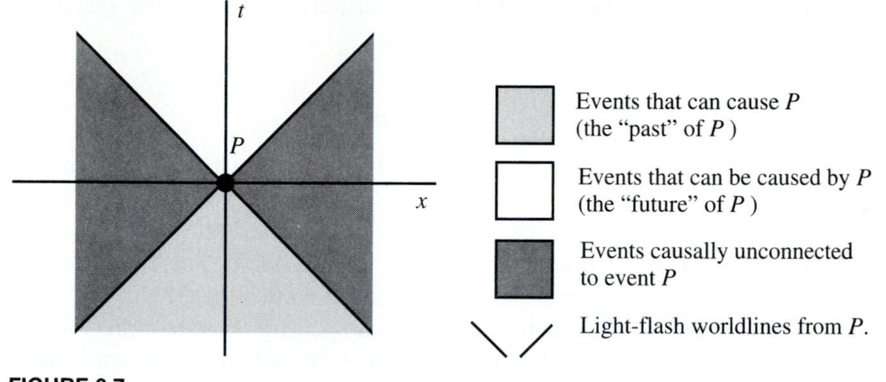

FIGURE 8.7
The causal structure of spacetime with regard to an event P.

into a still pool of water. If we plot the growth of such a ring on a spacetime diagram, we get a cone. The boundaries between the three regions described are then two tip-to-tip cones, as shown in Fig. 8.8. This boundary surface is often called the **light cone** for the given event P.

Figure 8.8 still leaves one spatial dimension unexpressed. The difficulty of representing (or even imagining) how the z dimension fits into such a diagram is obvious. This diagram represents the best that we can do in visualizing the four-dimensional reality of the light cone.

To summarize, the point of this section is that the spacetime interval classifications, which are basic, frame-independent features of the geometry of spacetime, have in fact a deeply physical significance: The sign of the squared spacetime interval between two events unambiguously describes whether these events can be causally connected or

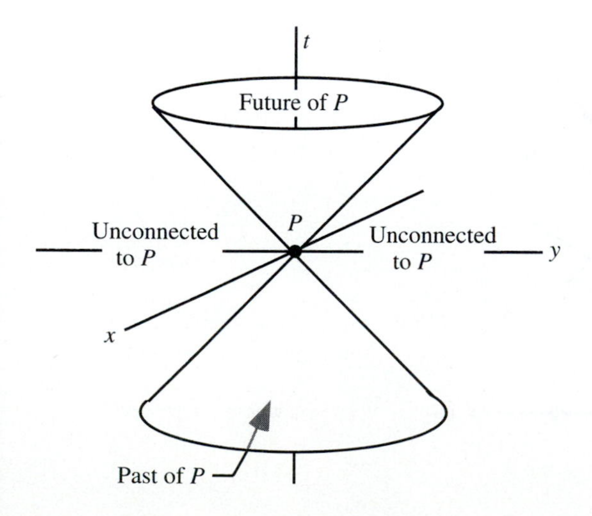

FIGURE 8.8
The light cone and regions of spacetime for the event P illustrated in a spacetime diagram with two spatial axes.

not. The light cone shown in Fig. 8.8 effectively illustrates this geometric feature of spacetime.

8.5 THE EINSTEIN VELOCITY TRANSFORMATION

In this section we turn our attention to the relativistic generalization of the galilean velocity transformation equations (1.3). Imagine a particle that is observed to move along the spatial x axis with a constant x velocity v'_x in the Other Frame, which itself is moving with a speed β in the positive x direction with respect to the Home Frame. What is the x velocity v_x of the particle as observed in the Home Frame?

Figure 8.9 shows how to construct a two-observer spacetime diagram that we can use to answer this question. After drawing both sets of coordinate axes, one simply draws the worldline of the particle in such a manner that its slope in the Other Frame is $1/v'_x$. The slope of the line in the Home Frame can then be easily determined by picking an arbitrary rise (5 ns in the diagram) and determining the run. One can then compute v_x by taking the inverse of the slope. In the case where $\beta = 3/5$ and $v'_x = 3/5$, Fig. 8.9 implies that the value of $v_x \approx 0.86$, *not* the value $3/5 + 3/5 = 6/5$ that the galilean velocity transformation equations predict.

Now let us see if we can derive an exact formula using the Lorentz transformation equations that does the same thing that this diagram does. We are trying to compute the x velocity v_x of a particle at a given instant of time as measured in the Home Frame knowing its x velocity v_x' as measured in the Other Frame, assuming that the frames

FIGURE 8.9
The particle shown here has an x velocity of $v_x' = 3/5$ as measured in the Other Frame, which itself has a velocity of $\beta = 3/5$ as measured in the Home Frame. The x velocity v_x of the particle in the Home Frame is seen from the diagram to be about 4.3 ns / 5 ns $\approx$ 0.86. Note that when $\beta = 3/5$, $\gamma = 5/4$.

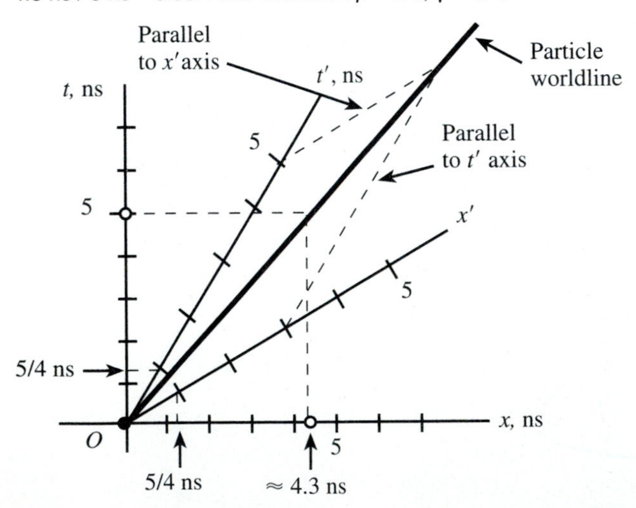

are in standard orientation. Pick two infinitesimally separated events along the worldline of the particle in question. Let the coordinate differences between these events as measured in the primary frame be dt and dx. Let the coordinate differences between the same two events as measured in the Other Frame be dt' and dx'. The x velocity of the particle as it travels between these events is

$$v_x \equiv \frac{dx}{dt} = \frac{\gamma(\beta\, dt' + dx')}{\gamma(dt' + \beta\, dx')} \tag{8.7}$$

where I have used the difference version of the inverse Lorentz transformation equations [Eqs. (6.14)]. Dividing the right side top and bottom by dt' and using $dx'/dt' = v_x'$, we get

$$v_x = \frac{\beta + v_x'}{1 + \beta v_x'} \qquad \text{(relativistically valid)} \tag{8.8a}$$

In a similar fashion, the y and z components of the particle's Home Frame velocity are found to be

$$v_y = \frac{v_y'\sqrt{1 - \beta^2}}{1 + \beta v_x'} \tag{8.8b}$$

$$v_z = \frac{v_z'\sqrt{1 - \beta^2}}{1 + \beta v_x'} \tag{8.8c}$$

As an example of the use of Eq. (8.8*a*), consider the particular problem illustrated by Fig. 8.9, where we have $v_x' = \beta = 3/5$. The final speed of the particle in the Home Frame is [according to Eq. (8.8*a*)]

$$v_x = \frac{3/5 + 3/5}{1 + (3/5)(3/5)} = \frac{6/5}{34/25} = \frac{30}{34} \approx 0.88 \tag{8.9}$$

This is in close agreement with the result read from the spacetime diagram in Fig. 8.9.

Equations (8.8) are called the **inverse Einstein velocity transformation equations** and express algebraically what Fig. 8.9 expresses graphically. Note that the result given here is different from what the galilean velocity transformation predicts: solving Eq. (1.3*a*) for v_x, we get

$$v_x = \beta + v_x' \qquad \text{(from the galilean transformation)} \tag{8.10}$$

Note that when the velocities β and v_x' are very small, the factor $\beta v_x'$ that appears in the denominator of Eq. (8.8*a*) becomes *very* small. In this limit the denominator $1 + \beta v_x' \approx 1$, and the Einstein equations reduce to the galilean equations:

$$v_x \approx \beta + v_x' \qquad \text{(in the low-velocity limit)} \tag{8.11}$$

The same kind of argument applies to the other two component equations as well. The galilean transformation equations are therefore reasonably accurate for everyday velocities but only represent an *approximation* to the true velocity transformation law expressed by Eqs. (8.8).

You should be able to convince yourself that Eq. (8.8a) never predicts that the x velocity of an object in the Home Frame will exceed the speed of light: even if both β and $v'_x = 1$ (which are their maximum possible values),

$$v_x = \frac{1+1}{1+1\cdot 1} = \frac{2}{2} = 1 \tag{8.12}$$

Moreover, this equation is consistent with the idea that the speed of light $= 1$ in all frames: if the x velocity of any object in the Other Frame is $v'_x = 1$, its speed in the Home Frame will be

$$v_x = \frac{\beta+1}{1+\beta\cdot 1} = \frac{\beta+1}{1+\beta} = 1 \tag{8.13}$$

independent of the value of β.

In short, the inverse Einstein velocity transformation equations [Eqs. (8.8)] provide the answer to the question that we raised in Chap. 2 about how the galilean velocity transformation equations can be modified to be consistent with the principle of relativity. Equations (8.8) reduce to the galilean transformation at low velocities (where the galilean transformation is known by experiment to be very accurate), but at relativistic velocities, they are consistent with both the assertion that nothing can be measured to go faster than light and the assertion that light itself has the same velocity in all inertial frames.

Equations (8.8) express the Home Frame velocity components v_x, v_y, and v_z in terms of the Other Frame velocity components v'_x, v'_y, and v'_z respectively. The direct **Einstein velocity transformation equations** express the Other Frame velocity components v'_x, v'_y, and v'_z in terms of the Home Frame velocity components v_x, v_y, and v_z, respectively:

$$v'_x = \frac{v_x - \beta}{1 - \beta v_x} \qquad v'_y = \frac{v_y\sqrt{1-\beta^2}}{1-\beta v_x} \qquad v'_z = \frac{v_z\sqrt{1-\beta^2}}{1-\beta v_x} \tag{8.14}$$

These equations can be derived either by solving Eqs. (8.8) for Other Frame quantities or by deriving them directly from the Lorentz transformation equations (6.13) in a manner similar to the way that Eqs. (8.8) were derived. (This is left as an exercise in Prob. 8.10.)

8.6 THE HEADLIGHT EFFECT (OPTIONAL)

One of the most interesting applications of these equations is the quantitative description of a relativistic phenomenon known as the *headlight effect*. Imagine a flashlight that in its own frame (which we will take to be the Other Frame) emits a beam of light in the shape of a cone whose sides make an angle of ϕ' with respect to the center of the beam, which (for simplicity) we will imagine to point in the $+x'$ direction. The situation is shown in Fig. 8.10*a*.

Now imagine that the flashlight and its beam are observed in a Home Frame in which the flashlight is moving with speed β in the $+x$ direction (i.e., in the direction of the center of its beam). In this frame, the Einstein velocity transformation equations (as we will see) imply that in this frame the sides of the beam's cone will be measured to make an angle ϕ with respect to the beam center that is *smaller* than the angle ϕ' measured in the flashlight's own frame. This means that in this frame, the beam is measured to be *narrower* than in the flashlight's own frame, as shown in Fig. 8.10*b*. This effective concentration of the light energy emitted by an object in the direction of its motion is aptly called the **headlight effect.**

Note that this effect is actually the *opposite* of what one might expect from a naive application of the idea of Lorentz contraction. One might expect that in the Home Frame the beam would be observed to be Lorentz-contracted in the direction of its motion, implying that the beam angle would be observed to be *larger* in this frame: the diameter of the beam will have the same value in both frames (since it is measured perpendicular to the line of relative motion), but its length will be contracted, making the angle larger.

This line of reasoning, however, is flawed because the beam is *not* a solid object (like the flashlight itself): it rather describes how the light emitted by the flashlight is observed to *move* in the Home Frame. The only way to find out how the emitted light is observed to *move* in the Home Frame compared to the flashlight's frame is to apply the Einstein velocity transformation equations. The Lorentz contraction equations are simply not relevant.

How do we apply the velocity transformation equations in this case? For the sake of simplicity, consider a particular ray of light that travels in the $x'y'$ plane (so that $v_z' = 0$) and also along the very edge of the beam's cone (i.e., so that the ray's angle with the center of the beam is ϕ' in the flashlight's frame). Such a ray is illustrated in Fig. 8.11*a*.

FIGURE 8.10
(*a*) A certain flashlight in its own reference frame (the Other Frame) is measured to a beam of light in the shape of a cone whose sides make an angle ϕ' with respect to the center of the beam. (*b*) In a frame where the flashlight moves in the direction of the beam with a speed equal to β (the Home Frame, the beam produced by the flashlight is observed to be *narrower:* the sides of the cone make an angle of $\phi < \phi'$ with respect to the center of the beam.

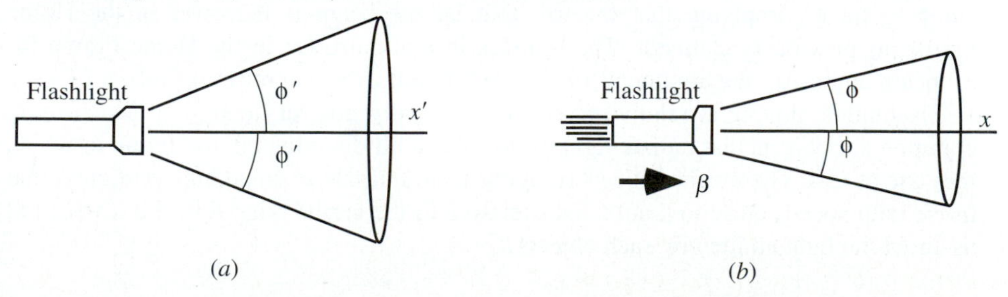

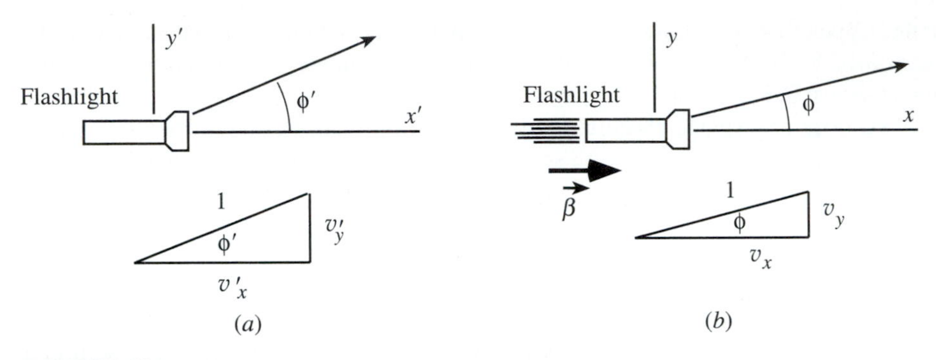

FIGURE 8.11
(*a*) In the Other Frame, a ray of light in the $x'y'$ plane traveling along the edge of the beam cone makes an angle ϕ' with the center of the beam. Since the speed of light is 1 in this frame, we can see from the triangle shown below the flashlight that $v_y' = \sin\phi'$ and $v_x' = \cos\phi'$. (*b*) In the Home Frame, this ray makes an angle of ϕ with respect to the center of the beam. Since the speed of light is 1 in this frame as well, we see from the triangle below the flashlight that $v_y = \sin\phi$ and $v_x = \cos\phi$.

Because the speed of light is 1 in both the flashlight's frame and the Home Frame, the diagrams shown in Fig. 8.11 tell us that

$$v_x' = \cos\phi' \qquad (8.15a)$$
$$v_y' = \sin\phi' \qquad (8.15b)$$

and

$$v_x = \cos\phi \qquad (8.15c)$$
$$v_y = \sin\phi \qquad (8.15d)$$

Plugging these results into the inverse Einstein velocity transformation equation for v_y [Eq. (8.8*b*)] we see that

$$\sin\phi = \frac{\sqrt{1-\beta^2}\sin\phi'}{1+\beta\cos\phi'} = \frac{\sin\phi'}{\gamma(1+\beta\cos\phi')} \qquad (8.16)$$

Since $\gamma > 1$ for nonzero β, it is easy to see that, for beam angles $\phi' \leq 90°$ at least, $\sin\phi < \sin\phi'$, implying that $\phi < \phi'$, that is, the beam is narrower in the Home Frame, as previously claimed. (The beam is in fact narrower in the Home Frame for *all* beam angles ϕ': the argument for $\phi' > 90°$ is left as an exercise in Prob. 8.23.)

Flashlights moving at relativistic speeds are really not very common. Much more common are objects that radiate light uniformly in all directions in the frame in which they are at rest. There exist objects (ranging from quasars to subatomic particles) that move with speeds close to that of light relative to the earth. What does Eq. (8.16) tell us about the light emitted by such objects?

If in its own frame an object emits light uniformly in all directions, then *half* the light that it emits will be emitted within a "cone" whose sides make an angle $\phi' = 90°$ with respect to the spatial x' axis in the object's own frame (see Fig. 8.12*a*). But according to Eq. (8.16), in a frame in which this object is moving with speed β, half the light produced by this object will be enclosed by a cone whose sides make an angle of

$$\phi = \sin^{-1}\frac{1}{\gamma} \tag{8.17}$$

(since $\phi' = 90°$ implies that $\sin\phi' = 1$ and $\cos\phi' = 0$): this is illustrated in Fig. 8.12*b*. For example, an object that radiates uniformly in its own frame will be seen in a frame in which it moves with a speed of 0.99 to emit half its light in a cone whose sides make an angle of

$$\phi = \sin^{-1}\frac{1}{\gamma} = \sin^{-1}\sqrt{1-\beta^2} = \sin^{-1}\sqrt{1-0.99^2} = \sin^{-1} 0.020 = 8.1° \tag{8.18}$$

with respect to the direction of the object's motion. You can see that the light emitted by an object that radiates uniformly in its own frame will be observed to be concentrated in the direction of its motion in any frame where it is moving. If the object is moving at a speed close to that of light, its light will be very strongly concentrated in that direction: this is why this effect is called the headlight effect.

This effect has been (indirectly) verified by observing the radiation produced by subatomic particles that have been accelerated to a large fraction of the speed of light in an accelerator. The radiation produced (called *synchrotron radiation*) has exactly the characteristics that one might expect if the radiation produced by each particle is subject to the headlight effect. Synchrotron radiation is also produced by certain astro-

FIGURE 8.12
(*a*) An object that emits light uniformly in the frame in which it is at rest (the Other Frame) will emit half of its light in the forward direction, i.e., within a "cone" whose sides make an angle of 90° with respect to the x' axis (the "cone" in this case is actually a flat disk that separates "forward" angles from "backward" angles). (*b*) In the Home Frame, where this object is traveling with a speed β, the same fraction of light is emitted inside a cone whose sides make an angle $\phi = \sin^{-1}(1/\gamma)$ with respect to the direction of its motion. The light from the object is therefore observed in this frame to be concentrated in the direction of motion. This effect becomes more pronounced as γ gets larger (i.e., as $\beta \to 1$).

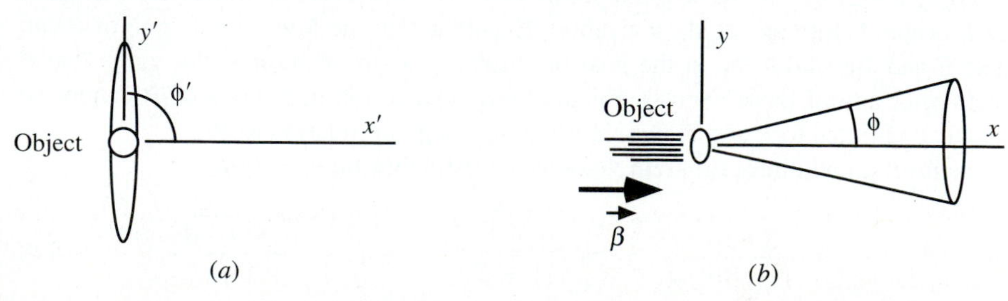

physical sources (such as pulsars), where energetic electrons are accelerated to relativistic speeds by intense electromagnetic fields. The point is that this effect has practical applications in both high-energy particle physics and astrophysics.

8.7 SUMMARY AND REVIEW

The most important ideas of this chapter can be summarized as follows.

The principle of relativity can only be made consistent with the concept of causality (which in a deep sense is an expression of the second law of thermodynamics) if whatever effect connects the causing event to the caused event travels between the events at *less* than (or at most equal to) the speed of light in a vacuum. This is because the caused event must follow the causing event in all inertial reference frames, and it is always possible to find a frame where the temporal order of two events is reversed if the coordinate separations of the event are such that $\Delta d > \Delta t$ in a given frame. Since any material object or indeed even just a message can be the agent of a cause-effect relationship, the speed of light represents an absolute upper limit to the speed that such an object or message can travel.

With the help of the Lorentz transformation equations, we saw that the squared spacetime interval Δs^2 between *any* two events is an invariant quantity, even when $\Delta d > \Delta t$ (making it impossible to have a clock present at both events). The interval between any two events can therefore be classified on the basis of the sign of $\Delta s^2 = \Delta t^2 - \Delta d^2$ as follows:

If $\Delta s^2 > 0$, the interval is **timelike,** and the events can be causally connected.
If $\Delta s^2 = 0$, the interval is **lightlike,** and the events can be causally connected.
If $\Delta s^2 < 0$, the interval is **spacelike,** and the events are causally *dis*connected.

Since the value of Δs^2 is frame-independent, observers in every reference frame will agree on the classification of the interval between any pair of events.

If the interval between two events is spacelike, the **spacetime separation** $\Delta\sigma \equiv \sqrt{\Delta d^2 - \Delta t^2}$ can be used to quantify the frame-independent interval between the events. This quantity is equal to the distance between the events as measured in a frame where the two events occur at the same time. Such a frame can be found for any events separated by a spacelike interval.

Spacetime has a frame-independent structure around a given event P expressed by that event's **light cone** (see Fig. 8.8). Events within the upper light cone of P can be caused by P and are said to be in the *future* of P (i.e., observers in all frames will agree that P occurs before any of these events). Events within the lower light cone of P can cause P and are said to be in the *past* of P (observers in all frames will agree that P occurs after any of these events). Events lying outside the light cone of P cannot be causally connected to P and have an ambiguous temporal relation to P.

The inverse and direct **Einstein velocity transformation equations**

$$v_x = \frac{\beta + v_x'}{1 + \beta v'} \qquad v_y = \frac{v_y'\sqrt{1-\beta^2}}{1 + \beta v_x'} \qquad v_z = \frac{v_z'\sqrt{1-\beta^2}}{1 + \beta v_x'} \tag{8.8}$$

and

$$v'_x = \frac{v_x - \beta}{1 - \beta v_x} \qquad v'_y = \frac{v_y\sqrt{1-\beta^2}}{1 - \beta v_x} \qquad v'_z = \frac{v_z\sqrt{1-\beta^2}}{1 - \beta v_x} \tag{8.14}$$

respectively, represent the relativistically correct generalizations of the galilean velocity transformation equations given by Eqs. (1.3) in Chap. 1. While these relativistic equations reduce to the galilean velocity transformations in the low-velocity limit, these equations are also consistent with the cosmic speed limit and the fact that light has the same velocity in all inertial frames.

The relativistic transformation of velocities can also be illustrated (and even quantitatively estimated) using a two-observer spacetime diagram, as shown in Fig. 8.9. The process is simple. Draw the worldline of the particle on the two-observer diagram to have the slope that correctly reflects the particle's x velocity in whichever frame you know that velocity, and then compute the slope of that worldline relative to the axes of the frame in which you are trying to find its velocity. This technique is entirely equivalent to using the Einstein velocity transformation equations, and if done carefully, can yield results accurate to better than 5 percent.

PROBLEMS

8.1 CAUSAL CLASSIFICATIONS. Which pairs of events in Fig. 8.13 could be causally connected? Which *cannot* be causally connected? Also, classify the spacetime interval between each pair as being spacelike, timelike, or lightlike, and state its value (or the value of the spacetime separation if necessary).

8.2 RELATIVISTIC DETECTIVE WORK. At 11:00:00 A.M. a boiler explodes in the basement of the Museum of Modern Art in New York City (call this event A). At 11:00:00.0003 A.M. a similar boiler explodes (call this event B) in the basement of a soup factory in Camden, New Jersey, a distance of 150 km from event A.

a Why is it impossible for the first event to have caused the second event?

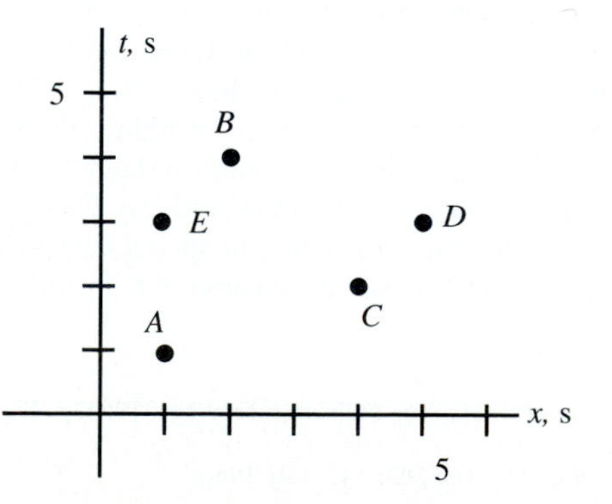

FIGURE 8.13

b An alien spaceship cruising in the direction of Camden from New York measures the Camden event to occur at the same time as the New York event. What is the approximate speed of the spaceship relative to the earth?

8.3 THE MEANING OF SIMULTANEITY.[1] "Joe Carter stubs his toe on a brick on earth. At the exact same instant, Knnnk Grblyx stubs her gnrrf on a zznkk on Alpha Centauri, which is 4.3 light-years from earth." The theory of relativity teaches us that this statement needs some kind of qualification before it becomes meaningful. Describe the qualification and why it is necessary.

8.4 THE COSMIC SPEED LIMIT. Here is a quick argument that no material object can go faster than the speed of light. Consider an object traveling in the $+x$ direction with respect to some inertial frame (call this the Home Frame) at a speed $v > 1$ in that frame. Show using a two-observer spacetime diagram that it is possible to find an Other Frame moving in the $+x$ direction at a speed $\beta < 1$ with respect to the Home Frame in which the worldline of the object lies along the diagram x' axis, and find the value of β (in terms of v) that makes this happen. Why is it absurd for the worldline of any object to coincide with the diagram x' axis?

8.5 FASTER THAN LIGHT.[2] Imagine spinning a laser at a speed of 100 rotations per second. How far away must you place a screen so that the spot of light on the screen produced by the laser beam sweeps along the screen at a speed faster than that of light? Would a spot speed greater than that of light violate the cosmic speed limit? (*Hint:* If you stand at one edge of the screen and a friend stands at the other, can you send a message to your friend using the sweeping spot? If so, how? If not, why not?)

8.6 SCISSORS PARADOX. Consider a very long pair of scissors. If you close the scissor blades fast enough, you might imagine that you could cause the *intersection* of the scissor blades (i.e., the point where they cut the paper) to travel from the near end of the scissors to the far end at a speed faster than that of light, without causing any *material part* of the scissors to exceed the speed of light. Argue that (1) this intersection can indeed travel faster than the speed of light in principle, but that (2) if the scissors are originally at open and at rest and you decide to send a *message* to a person at the other end of your scissors by suddenly closing them, you will find that the intersection *cannot* travel faster than the speed of light. Thus it is not possible to send a message faster than the speed of light using this technique.[3] (*Hint:* The information that the handles have begun to close must travel through the metal from the handles to the blades and then down the blades to cause the intersection to move forward. What effect carries this information? With what speed does this information travel?)

8.7 TRIP TO THE PAST. Starbase Alpha coasts in deep space at a speed of $\beta = 0.60$ in the $+x$ direction with respect to earth. Let the event of the starbase traveling by the earth define the origin event in both frames. Imagine that at $t = 8.0$ h, a giant accelerator on earth launches you toward the starbase at 10 times the speed of light, relative to the earth. After you get to the starbase, you use a similar accelerator to launch you back toward the earth at 10 times the speed of light, relative to the starbase. Using a carefully constructed (full-page) two-observer diagram, show that in such a case you will return to the earth before you left. (This is another way to illustrate the absurdity of faster-than-light travel.)

[1] Adapted from E. F. Taylor and J. A. Wheeler, *Spacetime Physics,* San Francisco: Freeman, 1966, p. 62.
[2] Ibid.
[3] See M. A. Rothman, "Things that Go Faster than Light," *Sci. Am.* **203:** 142, July 1960.

8.8 SPACELIKE OR TIMELIKE? Two balls are simultaneously ejected (event A) from the point $x = 0$ in some inertial frame. One rolls in the $+x$ direction with speed 0.80 and eventually runs into a wall at $x = 8.0$ ns (event B). The other rolls in the $-x$ direction with speed 0.40, eventually running into a wall at $x = -8.0$ ns (event C). Is the spacetime interval between B and C spacelike or timelike? Could these events be causally connected?

8.9 INTERVAL CLASSIFICATION. A flash of laser light is emitted by the earth (event A) and is absorbed on the moon (event B). Is the spacetime interval between events A and B spacelike, lightlike, or timelike? Now assume that the light flash is reflected at event B by something shiny on the moon and returns to earth, where it is absorbed at event C. Is the spacetime interval between B and C spacelike, lightlike, or timelike? What about the interval between events A and C? Support your answers by describing your reasoning.

8.10 THE DIRECT EINSTEIN VELOCITY TRANSFORMATION EQUATIONS. Derive Eqs. (8.14) two different ways:

a Solve Eqs. (8.8) for v'_x, v'_y, and v'_z.

b Use the Lorentz transformation equations (6.13) in a manner similar to the way that Eqs. (6.14) were used in the derivation of Eqs. (8.8).

8.11 VELOCITY TRANSFORMATION. A particle moves with velocity $v'_x = 2/5$ in an inertial frame attached to a train that in turn moves with velocity $\beta = 4/5$ in the $+x$ direction with respect to the ground. What is the velocity v_x of the particle with respect to the ground?

a Evaluate this by reading the velocity from a carefully constructed two-observer spacetime diagram.

b Check your answer using the appropriate Einstein velocity transformation equation.

8.12 VELOCITY TRANSFORMATION. Rocket A travels to the right and rocket B travels to the left at speeds of 3/5 and 4/5, respectively, relative to the earth. What is the velocity of A measured by observers in rocket B? Answer by reading the velocity from a carefully constructed two-observer diagram. Check your answer using the appropriate Einstein velocity transformation.

8.13 TRAIN WRECK. Two trains approach each other from opposite directions along a linear stretch of track. Each has a speed of 3/4 relative to the ground. What is the speed of one train relative to the other? Answer this question using an appropriate Einstein velocity transformation equation. (*Hint:* What frame should be chosen to be the Home Frame?)

8.14 RELATIVE VELOCITY. Two cars travel in the same direction on a Los Angeles freeway. Car A travels at a speed of 0.90, while car B can only muster a speed of 0.60. What is the relative speed of the cars?

8.15 GETTING FAST BY STAGES. The first stage of a multistage rocket boosts the rocket to a speed of 0.1 relative to the ground before being jettisoned. The next stage boosts the rocket to a speed of 0.1 relative to the final speed of the first stage, and so on. How many stages does it take to boost the payload to a speed in excess of 0.95?

8.16 PARTICLE INTERACTIONS IN TWO DIFFERENT FRAMES. Imagine that in the Home Frame two particles of equal mass m are observed to move along the x axis with equal and opposite speeds $v = 0.60$. The particles collide and stick together, becoming one big particle which remains at rest in the Home Frame. Now imagine observing the same situation from the vantage point of an Other Frame that moves with speed $\beta = 0.80$ in the $+x$ direction with respect to the Home Frame. Find the velocities of all the particles as observed in the Other Frame using the appropriate Einstein velocity transformation equations. Check your results using a two-observer spacetime diagram of the situation.

8.17 MOMENTUM NONCONSERVATION. In the situation described in Prob. 8.16, show that while the momentum of the system is conserved in the Home Frame, it is *not* conserved in the Other Frame if momentum is defined as mass times velocity. (We will deal with this problem in a future chapter).

8.18 VELOCITY TRANSFORMATION. A train travels in the $+x$ direction with a speed of $\beta = 0.80$ with respect to the ground. At a certain time, two balls are ejected, one traveling in the $+x$ direction with x velocity $v_{1x} = +0.60$ with respect to the train and one traveling in the $-x$ direction with x velocity $v_{2x} = -0.40$ with respect to the train.

a What are the x velocities of the balls with respect to the ground?
b What are the x velocities of the balls with respect to each other?

8.19 THE GALILEAN TRANSFORMATION EQUATIONS. Show that when $\beta << 1$, the Lorentz transformation equations (6.11*a*) and 6.11*b*) reduce to the galilean transformation equations (1.2*a*) and 1.2*b*). [*Discussion:* The problem is trivial except for Eq. (6.11*a*). In that equation, how can we justify dropping the βx term when we need to keep the $-\beta t$ term in (6.11*b*)? A typical distance coordinate found when measuring the motion of an object under everyday conditions would be on the order of several meters. How many seconds is this (order of magnitude)? A typical t coordinate would be on the order of a few seconds under the same conditions. With this in mind, does it make sense that $-\beta t$ should be kept when compared to x but βx can be neglected in comparison to t? Defend this position in your solution.]

8.20 THE VELOCITY OF LIGHT. Show using the Einstein velocity transformation equations (8.8) that a particle traveling in any arbitrary direction at the speed of light will be measured to have the speed of light in all other inertial frames.

8.21 VELOCITY ADDITION IN THREE DIMENSIONS. A particle moves with a speed of 4/5 in a direction 60° away from the spatial $+x'$ axis toward the spatial y' axis, as measured in a frame (the Other Frame) that is moving with speed $\beta = 0.50$ in the $+x$ direction of the Home Frame. What is the magnitude and direction of the particle's velocity in the Home Frame?

8.22 TWO-DIMENSIONAL VELOCITY ADDITION. A train travels in the $+x$ direction with a speed of $\beta = 0.80$ with respect to the ground. At a certain time, two balls are ejected so that they travel with a speed of 0.60 (as measured in the train frame) in opposite directions *perpendicular* to the direction of motion of the train.

a What are the speeds of the balls with respect to the ground?
b What is the angle that the paths of each ball makes with the x axis in the ground frame?

8.23 THE HEADLIGHT EFFECT. Show that Eqs. (8.15) and (8.8*a*) imply that

$$\cos\phi = \frac{\cos\phi' + \beta}{1 + \beta\cos\phi'} \tag{8.19}$$

Use this equation to argue that $\phi < \phi'$ for *all* ϕ'. (*Hint:* Show that $\cos\phi > \cos\phi'$.)

8.24 THE HEADLIGHT EFFECT. A flashlight is observed in its own frame to emit a beam in the shape of a cone whose sides make an angle of 30° with the beam's center. What is the beam angle in a frame where the flashlight is moving at a speed of 0.60 directly forward?

8.25 THE HEADLIGHT EFFECT. An object that radiates uniformly in its own frame is observed in the Home Frame to move with a speed of 0.80 in the $+x$ direction. A cone that encloses half the light emitted by the object makes what angle with respect to the x direction?

8.26 THE HEADLIGHT EFFECT. Imagine that a particle accelerator creates a beam of neutral muons traveling in the $+x$ direction with a speed of 0.9994. When neutral muons decay, they radiate gamma photons (high-energy particles of light) uniformly in all directions in their own frame. What cone will contain half these emitted photons in the laboratory frame?

8.27 THE HEADLIGHT EFFECT. An object that radiates uniformly in its own reference frame is observed in the Home Frame to move with a speed of 0.999 in the $+x$ direction.

a Argue that in the object's own frame, a (rather backward) "cone" whose sides make an angle of $\phi' = 168.5°$ with respect to the $+x$ direction encloses *approximately* 99 percent of the light emitted by the object.

b In the Home Frame, what cone will contain 99 percent of the light emitted by the object? [*Hint:* See Prob. 8.23. You will probably find Eq. (8.19) more useful than Eq. (8.16).]

8.28 THE EXTREME RELATIVISTIC LIMIT OF THE HEADLIGHT EFFECT. Show that for an object whose $\gamma >> 1$ (that is, whose speed is very close to the speed of light), the cone that encloses half the light emitted by the object makes an angle (in radians) of

$$\phi \approx \frac{1}{\gamma} \qquad \text{(when } \gamma >> 1\text{)} \tag{8.20}$$

with respect to its direction of motion.

9

FOUR-MOMENTUM

As soon as an equation seemed to him to be ugly, [Einstein] rather lost interest in it and could not understand why someone else was willing to spend much time on it. He was quite convinced that beauty was a guiding principle in the search for important results in theoretical physics.

H. Bondi[1]

9.1 OVERVIEW

Up to this point, we have been studying **relativistic kinematics** (*kinematics* is the study of how the motion of objects is to be measured and described mathematically.) But physicists are not only interested in *describing* motion, they are interested in explaining how objects interact and how the forces that they exert on each other *determine* their motion. The study of object interactions is called *dynamics.* In the remainder of this book, we will explore the basic principles of **relativistic dynamics.**

The basic principles of newtonian dynamics are expressed by Newton's famous three laws of motion. In Sec. 1.10 (see also Prob. 1.14 through 1.16), we saw that the laws of newtonian dynamics were consistent with the principle of relativity *if* the galilean velocity transformation equations [Eqs. (1.3)] are true. But as we saw in Chap. 8, the galilean velocity transformation equations are in fact *not* true; they only represent the low-velocity limit of the relativistically correct Einstein velocity transformations [Eqs. (8.14)]. *This means that the laws of newtonian dynamics are* not *gen-*

[1]Quoted in A. P. French (ed.), *Einstein: A Centenary Volume,* Cambridge, Mass.: Harvard, 1979, p.79.

erally consistent with the principle of relativity; the laws of newtonian dynamics likewise represent only low-velocity approximations to the laws of *relativistic* dynamics, laws that are the same in *all* inertial frames (as the principle of relativity requires).

Well, what are these laws of relativistic dynamics, and how can we find them? We *could* address this question by searching for a relativistic generalization of Newton's second law, then Newton's third law, then the law of universal gravitation, and so on. This approach seems on the surface to be logical and straightforward. But it turns out that the most fruitful way to address this problem is to focus our attention on the law of conservation of momentum. Not only is the correct relativistic generalization of this law fairly easy to find, but this law proves to be extremely rich in implications and applications. Indeed, virtually everything that is useful to know about relativistic dynamics can be learned by a close examination of the law of conservation of momentum.

The basic argument in this chapter can be outlined as follows.

1. First, I will show you that conservation of ordinary newtonian momentum (defined as mass × velocity) is *not* consistent with the principle of relativity: if the total newtonian momentum of a system of interacting objects is conserved in one inertial reference frame, it will *not* be conserved in other inertial frames. This means that we need to look for an appropriate relativistic modification of the idea of momentum that *can* be made consistent with the principle of relativity.
2. I will then propose a natural relativistic generalization of the idea of momentum called the *four-momentum.* This four-momentum is a *four*-component vector having a *time component* as well as the usual x, y, and z components.
3. I will then show you that a law of conservation of four-momentum *is* consistent with the principle of relativity (and thus represents a reasonable candidate as a relativistic generalization of the newtonian law of conservation of momentum). But for this to be true, the t component of a system's total four-momentum vector *must be conserved* along with its x, y, and z components. So the law of conservation of four-momentum not only generalizes the idea of conservation of newtonian momentum, it tells us that *something else is conserved as well.*
4. This fourth conserved quantity (i.e., the t component of the four-momentum vector) turns out to be a relativistic generalization of the idea of *energy.* Thus the law of conservation of four-momentum not only represents a relativistic generalization of the newtonian law of conservation of momentum but the newtonian law of conservation of energy as well! Special relativity thus ends up giving us two conservation laws for the price of one!

Let us see how this all works out.

9.2 NEWTONIAN MOMENTUM IS NOT CONSERVED IN ALL FRAMES

The law of conservation of momentum must satisfy the principle of relativity if it is to be a valid physical law. This implies that if the total momentum of a system of interacting particles is conserved in *one* inertial reference frame (and the law of conservation of momentum is a valid physical law), then it should be conserved in *all* inertial reference frames.

In this section, I will show you that the law of conservation of newtonian momentum is *not* consistent with the principle of relativity and thus cannot be a valid law of physics. To illustrate the problem, it is sufficient to demonstrate a single instance of the inconsistency. For the sake of simplicity, I will illustrate the difficulty with the ordinary newtonian definition of momentum using a simple one-dimensional collision.

Figure 9.1 shows such a collision as viewed in the Home Frame. In this frame, an object of mass m is moving in the $+x$ direction with an x velocity $v_{1x} = +3/4$. It then strikes an object of mass $2m$ at rest ($v_{2x} = 0$). The lighter mass rebounds from the collision with an x velocity of $v_{3x} = -1/4$, while the heavier object rebounds with an x velocity of $v_{4x} = +1/2$. The system can be considered to be isolated from external effects.

The system's total newtonian x momentum is conserved in the Home Frame in this case:

$$\text{Total } x \text{ momentum } \textit{before}: \quad mv_{1x} + 2mv_{2x} = m(+3/4) + 2m(0) = +3m/4 \tag{9.1a}$$

$$\text{Total } x \text{ momentum } \textit{after}: \quad mv_{3x} + 2mv_{4x} = m(+1/4) + 2m(+1/2) = +3m/4 \tag{9.1b}$$

Now consider how this collision looks when observed in an Other Frame that is moving with a speed $\beta = 3/4$ in the $+x$ direction. Since this frame essentially moves along with the lightweight particle, it appears to be at rest in the Other Frame: $v'_{1x} = 0$. Since the larger object is at rest in the Home Frame and the Home Frame appears to be moving backward with respect to the Other Frame at a speed of 3/4, the x velocity of the more massive object must be $v'_{2x} = -3/4$ as well. In this frame, then, the collision will be measured to happen as shown in Fig. 9.2.

FIGURE 9.1
A hypothetical collision of two particles as viewed in the Home Frame. The total newtonian momentum of this system is conserved in this frame.

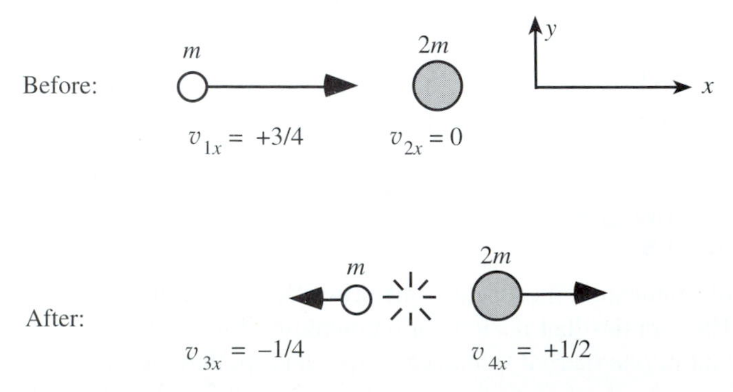

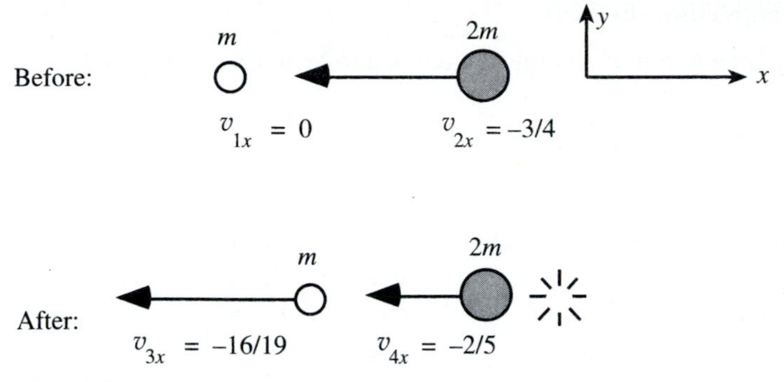

FIGURE 9.2
The same collision process as observed in the Other Frame. Newtonian momentum is *not* conserved in this frame.

The objects' final x velocities are not so easy to intuit: we need to compute these velocities using the first of the Einstein velocity transformation equations (8.14):

$$v'_{3x} = \frac{v_{3x} - \beta}{1 - \beta v_{3x}} = \frac{-1/4 - 3/4}{1 - (3/4)(-1/4)} = -\frac{16}{19} \tag{9.2a}$$

$$v'_{4x} = \frac{v_{4x} - \beta}{1 - \beta v_{4x}} = \frac{+1/2 - 3/4}{1 - (3/4)(1/2)} = -\frac{2}{5} \tag{9.2b}$$

Note that these *must* be the final velocities of the two objects if the Einstein velocity transformation equation is true and the collision actually takes place as shown in Fig. 9.2.

The system's total newtonian x momentum is *not* conserved in this case!

Total x momentum *before:* $\quad mv'_{1x} + 2mv'_{2x} = m(0) + 2m(-3/4) = -3m/2 \qquad$ (9.3a)

Total x momentum *after:* $\quad mv'_{3x} + 2mv'_{4x} = m(-2/5) + 2m(-16/19)$
$= -156m/95 \qquad$ (9.3b)

We can see in this frame that the x component of the momentum is somewhat larger after the collision than it was before the collision, since $156/95 > 3/2$. The law of conservation of newtonian momentum therefore does *not* hold in the Other Frame, even though it did hold in the Home Frame. (Note that the law of conservation of momentum requires that *each* component of the system's total momentum be conserved separately, so a violation of the conservation of even *one* component, the x component in this case, is a violation of the entire law.)

The principle of relativity requires that the laws of physics be the same in all inertial reference frames. The conclusion is inescapable: If the Einstein velocity transformation equations are true, then the law of conservation of newtonian momentum is *not* consistent with the principle of relativity.

9.3 THE FOUR-MOMENTUM VECTOR

Ordinary newtonian momentum $\vec{p}$ of an object is defined to be mass times velocity:

$$\vec{p} \equiv m\vec{v} = m\,\frac{d\vec{r}}{dt} \tag{9.4}$$

The vector $d\vec{r}$ represents an infinitesimal displacement in space, which is divided by an infinitesimal time dt to get the object's velocity vector $\vec{v}$. The components of the infinitesimal displacement vector $d\vec{r}$ are $[dx, dy, dz]$, so the components of the newtonian momentum are

$$p_x = mv_x = m\,\frac{dx}{dt} \qquad p_y = m\,\frac{dy}{dt} \qquad p_z = m\,\frac{dz}{dt} \tag{9.5}$$

Notice that the momentum vector is parallel to the infinitesimal displacement $d\vec{r}$ and so will be tangent to the path of the object through space, as shown in Fig. 9.3.

How can we arrive at a relativistic generalization of this process? In special relativity, space and time are considered to be equal parts of a whole called *spacetime;* thus we describe the motion of an object not merely by describing its path through space but by describing its *worldline through spacetime.* The appropriate relativistic generalization of an infinitesimal displacement in space $d\vec{r}$ between two infinitesimally separated points on an object's path in space is a displacement $d\mathbf{R}$ in *spacetime* between two infinitesimally separated *events* on the object's *worldline* (see Fig. 9.4). Note that the displacement $d\mathbf{R}$ in spacetime between two events is specified by *four* numbers:

$$d\mathbf{R} = [dt, dx, dy, dz] \tag{9.6}$$

Including the time displacement dt on an equal footing with the spatial displacements dx, dy, dz makes the displacement $d\mathbf{R}$ a four-component vector called a **four-vector.** In this book, I will always use *boldface capital letters* ($d\mathbf{R}$, $\mathbf{P}$, etc.) to represent such four-vectors.

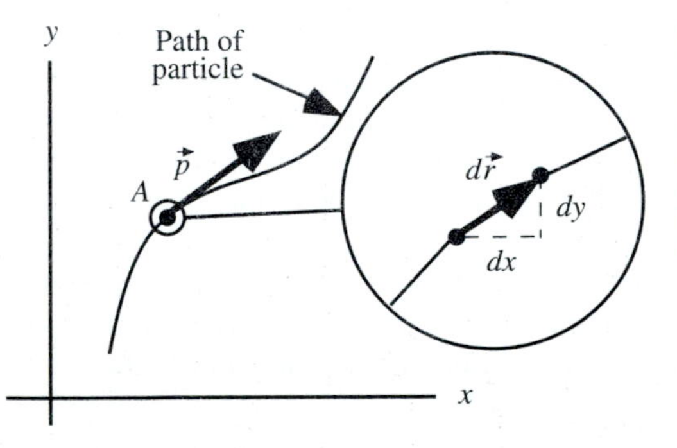

FIGURE 9.3
The path of a particle through space is shown in the graph at left. The ordinary momentum $\vec{p}$ of the particle at point A is defined to be a vector parallel to the displacement vector $d\vec{r}$ that connects two infinitesimally separated points surrounding A.

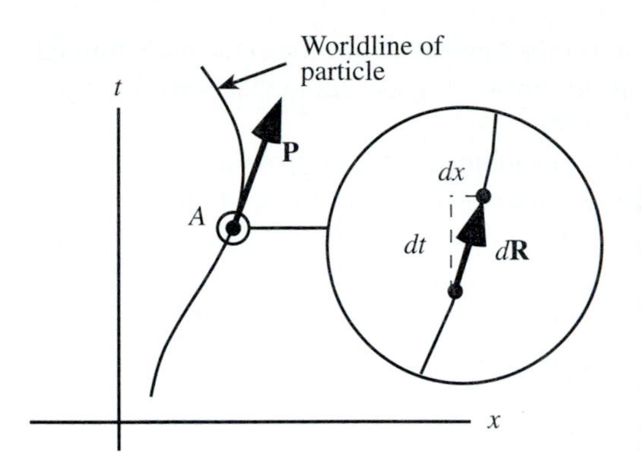

FIGURE 9.4
The worldline of an object through space is shown in the spacetime diagram at left. The relativistic momentum **P** of the object at event A is defined to be an arrow parallel (on the diagram) to the displacement arrow $d\mathbf{R}$ that connects two infinitesimally separated events surrounding the event A. (The y and z dimensions are not shown on this diagram.)

Given the components $[dt, dx, dy, dz]$ of the displacement four-vector $d\mathbf{R}$ that stretches between two infinitesimally separated events on the worldline of our object, how do we define its relativistic momentum? By analogy with the newtonian momentum, we want to divide $d\mathbf{R}$ by a quantity that somehow represents the *time* between the two events and then multiply by the mass of the object. But in the theory of relativity, the time measured between two events depends on who does the measuring! Whose time shall we use?

There is *one* kind of time that can be measured between the events that has a frame-independent value and is uniquely related to the motion of the object in question: this time is the infinitesimal proper time $d\tau$ between the events that would be measured by a clock traveling *with the object itself.* When you think about it, it makes a certain kind of sense to use the object's *own* time to characterize its momentum. Therefore our proposed relativistic generalization of the newtonian momentum $\vec{p}$ of an object is the relativistic four-momentum **P,** defined as follows:

$$\mathbf{P} = m \frac{d\mathbf{R}}{d\tau} \tag{9.7}$$

The four-momentum is a four-dimensional vector having components $[P_t, P_x, P_y, P_z]$ such that

$$P_t = m \frac{dt}{d\tau} \tag{9.8a}$$

$$P_x = m \frac{dx}{d\tau} \tag{9.8b}$$

$$P_y = m \frac{dy}{d\tau} \tag{9.8c}$$

$$P_z = m \frac{dz}{d\tau} \tag{9.8d}$$

Just as the newtonian momentum is a vector tangent to the object's *path* through space, the four-momentum (displayed as an arrow on a spacetime diagram) is tangent to the object's *worldline,* as shown in Fig. 9.4.

We can express the components of the four-momentum in a given inertial frame in terms of the ordinary velocity of the object measured in that frame. Equation (5.11) tells us that the infinitesimal proper time between two infinitesimally separated events measured by a clock traveling between two events at a speed v in a given reference frame is related to the coordinate time dt measured between those events by

$$d\tau = \sqrt{1 - v^2}\, dt \tag{9.9}$$

This means that

$$P_t \equiv m\frac{dt}{d\tau} = \frac{m}{\sqrt{1 - v^2}}\frac{dt}{dt} = \frac{m}{\sqrt{1 - v^2}} \tag{9.10a}$$

$$P_x \equiv m\frac{dx}{d\tau} = \frac{m}{\sqrt{1 - v^2}}\frac{dx}{dt} = \frac{mv_x}{\sqrt{1 - v^2}} \tag{9.10b}$$

$$P_y \equiv m\frac{dy}{d\tau} = \frac{m}{\sqrt{1 - v^2}}\frac{dy}{dt} = \frac{mv_y}{\sqrt{1 - v^2}} \tag{9.10c}$$

$$P_z \equiv m\frac{dz}{d\tau} = \frac{m}{\sqrt{1 - v^2}}\frac{dz}{dt} = \frac{mv_z}{\sqrt{1 - v^2}} \tag{9.10d}$$

These equations allow one to find the components of the four-momentum of an object in a given frame knowing the object's velocity vector $\vec{v}$ in that frame.

When the speed v of an object becomes very small compared to the speed of light ($v << 1$), the square roots in the denominators in Eqs. (9.10) become almost equal to 1, and we have

$$P_t \approx m \tag{9.11a}$$
$$P_x \approx mv_x \quad \text{(when } v << 1\text{)} \tag{9.11b}$$
$$P_y \approx mv_y \tag{9.11c}$$
$$P_z \approx mv_z \tag{9.11d}$$

Thus in the limit where the velocity of an object is very small (i.e., ordinary, everyday velocities), the spatial components of the four-momentum reduce to being the same as the corresponding components of the object's newtonian momentum.

Note also that since velocity in the SR unit system is a unitless number, all four components of an object's four-momentum have units of *mass* in the SR system.

9.4 PROPERTIES OF THE FOUR-MOMENTUM VECTOR

Why define an object's relativistic four-momentum this way? The definition has several attractive features. One feature has already been mentioned: *on a spacetime dia-*

gram, the four-momentum is represented by an arrow tangent to the worldline of the object through spacetime, just as the ordinary momentum vector is represented by an arrow tangent to the path of the object through space.

It is also nice that the definition of the four-momentum treats the time coordinate in the same manner as the spatial coordinates: all four coordinate displacements dt, dx, dy, dz appear on an equal footing in the definition of the four-momentum given by Eqs. (9.8). We have already seen how it is important in relativity theory to treat time and space as being equal participants in the larger geometric whole that we call spacetime. The definition of the four-momentum given above maintains this symmetry between the different spacetime coordinates.

All this symmetry has a certain beauty about it which a physicist like Einstein might take as corroborating evidence that we are on the right track with this definition. But we will see that the most important feature of the definition is that, given the components of the four-momentum in a particular reference frame, there is a very simple and straightforward method for calculating what its components will be measured to be in any *other* inertial reference frame. Let us find out what this transformation rule is.

The components of the four-momentum of an object are *frame-dependent* quantities, because the values of the coordinate differences dt, dx, dy, dz that appear in the numerators of Eqs. (9.8) are frame-dependent. The differential proper time $d\tau$ appearing in the denominator, on the other hand, is a *frame-independent* quantity. In this text, we will also consider the mass m of the object to be a *frame-independent* measure of the amount of "stuff" in the object.[1]

Let us imagine that we know the components $[P_t, P_x, P_y, P_z]$ of a given object's four-momentum in the Home Frame and that we want to find the corresponding components in an Other Frame moving with speed β in the $+x$ direction with respect to the Home Frame. The time component P_t' of the object's four-momentum in the Other Frame can be calculated as follows:

$$P_t' = m\frac{dt'}{d\tau} = m\frac{\gamma(dt - \beta\, dx)}{d\tau} = \gamma m\frac{dt}{d\tau} - \gamma\beta m\frac{dx}{d\tau}$$
$$\Rightarrow \gamma(P_t - \beta P_x) \tag{9.12a}$$

where I have used the Lorentz transformation [specifically Eq. (6.13*a*)] to express dt' as measured in the Other Frame in terms of dt and dx as measured in the Home Frame. Similarly, the transformation equation for the four-momentum x component is (as *you* can show)

$$P_x' = \gamma(-\beta P_t + P_x) \tag{9.12b}$$

[1]You may have heard in another context that special relativity implies that the mass of an object depends on its velocity. This is an old-fashioned way of looking at mass that obscures some of the simplicity and beauty of relativity theory. See C. G. Adler, "Does Mass Really Depend on Velocity, Dad?" *Am. J. Phys.*, **55**(8): 739–743, August 1987, for a careful and entertaining look at the problems with the old way of thinking about mass in relativity theory. Most modern treatments of relativity treat mass as being frame-independent.

We also have

$$P'_y = m\frac{dy'}{d\tau} = m\frac{dy}{d\tau} = P_y \tag{9.12c}$$

$$P'_z = m\frac{dz'}{d\tau} = m\frac{dz}{d\tau} = P_z \tag{9.12d}$$

Compare these with the Lorentz transformation equations (6.13). Note that Eqs. (9.12) are exactly the same as the Lorentz transformation equations except that the four-momentum components P_t, P_x, P_y, P_z, P'_t, P'_x, P'_y, and P'_z have been substituted for the coordinate displacement components Δt, Δx, Δy, Δz, $\Delta t'$, $\Delta x'$, $\Delta y'$, and $\Delta z'$, respectively. *Thus the components of the four-momentum transform from frame to frame according to the Lorentz transformation equations, just as coordinate differences do!*

The transformation equations for the four-momentum come out so nicely because (1) the time coordinate appears on an equal footing with the spatial components in the definition of the four-momentum, (2) we have divided the displacement by the frame-independent differential proper time $d\tau$ instead of the frame-dependent differential coordinate time dt, and (3) we are considering the mass m to be a frame-independent quantity.

In fact, the technical definition of a four-vector requires this kind of transformation law:

Definition of a Four-Vector

A **four-vector** is a physical quantity represented by a vector having four components that transform according to the Lorentz transformation equations (i.e., just as the coordinate differences Δt, Δx, Δy, Δz do) when we go from one inertial reference frame to another.

Now we know that although the coordinate differences Δt, Δx, Δy, Δz between two events are frame-dependent quantities, the spacetime interval Δs between the events given by

$$\Delta s^2 = \Delta t^2 - \Delta x^2 - \Delta y^2 - \Delta z^2 \tag{9.13}$$

is a frame-*in*dependent quantity. Similarly, we can define a frame-independent number called the **four-magnitude** of a four-vector as follows: If a four-vector $\mathbf{A}$ has components $[A_t, A_x, A_y, A_z]$, then its squared four-magnitude is defined to be

$$|\mathbf{A}|^2 = A_t^2 - A_x^2 - A_y^2 - A_z^2 \tag{9.14}$$

This is analogous to using the pythagorean theorem to find the magnitude of an ordinary vector.

You can easily show that the four-magnitude of an object's four-momentum vector is equal to its frame-independent mass:

$$m = |\mathbf{P}| = \sqrt{P_t^2 - P_x^2 - P_y^2 - P_z^2} \tag{9.15}$$

The easiest way to do this involves using the fact that for infinitesimally separated events, there is no distinction between proper time and spacetime interval: $d\tau = ds = \sqrt{dt^2 - dx^2 - dy^2 - dz^2}$. Alternatively, you can evaluate the square root in Eq. (9.15) directly using Eqs. (9.11). See if you can verify this very important and useful result now.

9.5 FOUR-MOMENTUM CAN BE CONSERVED IN *ALL* REFERENCE FRAMES

We now have a suitable candidate for a relativistic generalization of the concept of momentum. The final touch is to verify that a law of conservation of four-momentum is in fact consistent with the principle of relativity. It is important to check this (this is where the law of conservation of newtonian momentum fails!).

Consider an arbitrary collision of two objects moving along the x axis. The law of conservation of four-momentum says that

$$\mathbf{P}_1 + \mathbf{P}_2 = \mathbf{P}_3 + \mathbf{P}_4 \tag{9.16}$$

where $\mathbf{P}_1$ and $\mathbf{P}_2$ are the objects' four-momenta *before* the collision and $\mathbf{P}_3$ and $\mathbf{P}_4$ are the objects' four-momenta *after* the collision. This equation can be fruitfully rewritten in the form

$$\mathbf{P}_1 + \mathbf{P}_2 - \mathbf{P}_3 - \mathbf{P}_4 = 0 \tag{9.17}$$

which essentially says that the *difference* between the system's initial and final total momenta is zero. In component form, this last equation tells us that

$$\begin{bmatrix} P_{1t} \\ P_{1x} \\ P_{1y} \\ P_{1z} \end{bmatrix} + \begin{bmatrix} P_{2t} \\ P_{2x} \\ P_{2y} \\ P_{2z} \end{bmatrix} - \begin{bmatrix} P_{3t} \\ P_{3x} \\ P_{3y} \\ P_{3z} \end{bmatrix} - \begin{bmatrix} P_{4t} \\ P_{4x} \\ P_{4y} \\ P_{4z} \end{bmatrix} = \begin{bmatrix} 0 \\ 0 \\ 0 \\ 0 \end{bmatrix} \quad \text{or} \quad \begin{aligned} P_{1t} + P_{2t} - P_{3t} - P_{4t} &= 0 \\ P_{1x} + P_{2x} - P_{3x} - P_{4x} &= 0 \\ P_{1y} + P_{2y} - P_{3y} - P_{4y} &= 0 \\ P_{1z} + P_{2z} - P_{3z} - P_{4z} &= 0 \end{aligned} \tag{9.18}$$

When expressed in component form, the single equation (9.17) becomes a set of *four* equations, one for each of the four components of the four-momentum. Each of these equations has to be independently satisfied if four-momentum is to be conserved.

Let us assume that we have observed a collision in the Home Frame and we have determined it satisfies the law of conservation of four-momentum in that frame. The principle of relativity requires that the same law apply in *every other* inertial reference frame, i.e.,

$$\begin{aligned} &\text{If} \quad \mathbf{P}_1 + \mathbf{P}_2 - \mathbf{P}_3 - \mathbf{P}_4 = 0 \quad \text{in the Home Frame} \\ &\text{Then} \quad \mathbf{P}_1' + \mathbf{P}_2' - \mathbf{P}_3' - \mathbf{P}_4' = 0 \quad \text{in any Other Frame} \end{aligned} \tag{9.19}$$

If this statement is not true, our proposed relativistic generalization of the idea of momentum is not any better than newtonian momentum. If the statement *is* true, then the law of conservation of four-momentum represents at least a *possible* relativistic generalization of the law of conservation of ordinary momentum.

We can in fact easily show that this is true for any collision as viewed in any Other inertial reference frame. Consider the x component of the conservation law in the Other Frame. According to the transformation law for the components of the four-momentum given by Eq. (9.12*b*),

$$P'_{1x} + P'_{2x} - P'_{3x} - P'_{4x} = \gamma(-\beta P_{1t} + P_{1x}) + \gamma(-\beta P_{2t} + P_{2x}) \\ -\gamma(-\beta P_{3t} + P_{3x}) - \gamma(-\beta P_{3t} + P_{3x}) \quad (9.20a)$$

Collecting the terms on the right that are multiplied by γ and those multiplied by $\gamma\beta$, we get

$$P'_{1x} + P'_{2x} - P'_{3x} - P'_{4x} = -\gamma\beta(P_{1t} + P_{2t} - P_{3t} - P_{4t}) \\ +\gamma(P_{1x} + P_{2x} - P_{3x} - P_{4x}) \quad (9.20b)$$

But if both the t and x components of the four-momentum are conserved in the Home Frame, then the first two lines of Eq. (9.18) tell us that the quantities in parentheses equal zero. So

$$P'_{1x} + P'_{2x} - P'_{3x} - P'_{4x} = -\gamma\beta(0) + \gamma(0) = 0 \quad (9.20c)$$

meaning that *if* both the t and x components of the system's total four-momentum are conserved in the Home Frame, then the x component of the system's total four-momentum will also be observed to be conserved in the Other Frame, as hoped. The proof that the *other* components of the system's total four-momentum are *also* conserved in the Other Frame is essentially the same (check this!).

What we have shown is that the law of conservation of four-momentum expressed by Eq. (9.17) is *consistent* with the principle of relativity in the sense that if it holds in one frame it holds in all. That does not make the law *true:* it simply makes it *possible.* But now let me argue for the *truth* of this law. (1) We know from a multitude of experiments at low velocities that *some* quantity that reduces to newtonian momentum at such velocities is conserved. (2) Conservation of newtonian momentum will not work: the law is inconsistent with the principle of relativity. (3) The hypothetical law of conservation of four-momentum *is* compatible with the principle of relativity. (4) The three spatial components of the four-momentum *do* reduce to the components of newtonian momentum at low velocities. (5) Therefore, if the law of conservation of four-momentum were true, both would explain the low-velocity experimental data and maintain compatibility with the principle of relativity. In short, our observations at low velocities suggest that *something* like momentum is conserved, and in the absence of compelling alternatives, it only makes *sense* to believe that it is the total four-momentum of a system that is conserved.

Of course, no matter how suggestive a theoretical argument might be, there is no substitute for direct experimental evidence. Since the 1950s, physicists have been using particle accelerators to create beams of subatomic particles traveling at speeds very near the speed of light and colliding these particles with stationary targets or other particle beams. At such speeds, the distinction between newtonian momentum and four-momentum is very clear, and analysis of a typical experiment involves applying conservation of four-momentum to anywhere from thousands to millions of particle collisions. The result is that conservation of four-momentum is implicitly tested thousands of times daily in the course of such research.[1] In spite of this enormous wealth of data, no compelling evidence of a violation of the law of conservation of four-momentum has ever been seen.

9.6 THE TIME COMPONENT OF FOUR-MOMENTUM

Equations (9.20*b*) and (9.20*c*) make it clear that conservation of four-momentum is *only* consistent with the principle of relativity if all *four* components of the four-momentum (P_t as well as P_x, P_y, and P_z) are independently conserved. The three spatial components of an object's four-momentum correspond (at low velocities) to the three components of its newtonian momentum. But the law of conservation of four-momentum requires that something else (that is, P_t) be conserved as well. What is the physical interpretation of this new conserved quantity?

According to Eq. (9.10*a*), the time component of an object's four-momentum is:

$$P_t = \frac{m}{\sqrt{1 - v^2}} \qquad (9.21)$$

We know that this reduces to the mass m of the object at low velocities but is not exactly equal to the mass. Let us use the binomial approximation [Eq. (5.4)] to find out how this quantity differs from m when $v << 1$. Since $(1 - x)^a \approx 1 - ax$ if $x << 1$, we have

$$P_t = \frac{m}{\sqrt{1 - v^2}} = m(1 - v^2)^{-1/2} \approx m[1 - (-\tfrac{1}{2})v^2] = m + \tfrac{1}{2} mv^2 \qquad (9.22)$$

when $v << 1$.

The first term here is the mass of the particle, and when v is zero, that is what P_t becomes. But when v is nonzero but still very small, we have an additional term in P_t that corresponds to the *kinetic energy* of the particle. If P_t is conserved in a collision at low velocities, what we are saying is that the sum of the masses plus the sum of the

[1]E. F. Taylor and J. A. Wheeler, *Spacetime Physics,* San Francisco: Freeman, 1966, estimated at that time that as a by-product of particle physics research the law of conservation of four-momentum was (implicitly) checked more frequently than axioms of euclidean geometry were (implicitly) checked by surveyors making measurements of the surface of the earth!

kinetic energies of the particles involved in the collision are conserved. If the masses of the particles remain unchanged in the collision, then conservation of P_t tells us that the kinetic energy of the particles is conserved.

Thus the statement about the conservation of P_t is (at low velocities) basically the newtonian statement of conservation of [kinetic] energy! As we have generalized the concept of momentum, so now we will generalize the concept of energy. We *define* the component P_t of a particle's four-momentum to be the particle's **relativistic energy** E and assert that it is this relativistic energy that is the fourth quantity conserved in an isolated system. The relativistic energy E of an object moving at a speed v as measured in a given reference frame is thus defined to be

$$E \equiv P_t = \frac{m}{\sqrt{1 - v^2}} \tag{9.23}$$

Note that if the speed v of the object is an appreciable fraction of the speed of light, the approximation given by Eq. (9.22) does not hold. The *relativistic kinetic energy* of an object is defined (for all v) to be the difference between the object's total relativistic energy E and its mass-energy m:

$$K \equiv E - m = \frac{m}{\sqrt{1 - v^2}} - m = m\left(\frac{1}{\sqrt{1 - v^2}} - 1\right) \tag{9.24}$$

The relativistic kinetic energy K becomes approximately equal to $\frac{1}{2}mv^2$ only for small values of v: in general, $K > \frac{1}{2}mv^2$.

In newtonian mechanics, conservation of energy and momentum were thought of as separate concepts. But just as special relativity binds space and time into a single geometry, so here it binds the laws of conservation momentum and energy into a single statement: *The total four-momentum of an isolated system of particles is conserved.* Conservation of energy is impossible without conservation of momentum, and vice versa: energy and momentum are indissolvably bound together as parts of the same whole.

Note, however, that the relativistic energy of an object is not just the kinetic energy of that object (even at low velocities) but includes the *mass* of the object as well. The fact that $E = P_t$ is conserved does *not* imply that the mass of the object and its kinetic energy are *separately* conserved, only that the *whole* (i.e., the relativistic energy) is conserved. This implies that *processes that convert mass to kinetic energy, and vice versa, do not necessarily violate the law of conservation of four-momentum* and therefore might exist. This subject will be more fully explored in the next chapter.

Equation (9.22) is expressed in SR units, where velocity is unitless and both kinetic energy and mass are measured in kilograms. If we would like to express the relativistic energy of the particle in the SI unit of joules = kg·m^2/s^2, we must multiply the energy in kilograms by two powers of the conversion factor $c = 2.998 \times 10^8$ m/s to get the units to come out right. Therefore, Eq. (9.22) in SI units would read

$$E \approx mc^2 + \frac{1}{2}mv^2 \qquad \text{when } v^2 << c^2 \text{ (SI units)} \tag{9.25}$$

where the energy E is measured in joules and the speed v is measured in meters per second. In particular, when the particle is at rest, its relativistic energy in SI units is

$$E_{\text{rest}} = mc^2 \tag{9.26}$$

This is the famous equation that has served as an icon representing both the essence of special relativity and also Einstein's achievement. It should be recognized that this equation is simply a special case of the more general Eq. (9.23). Nonetheless, it does focus our attention on the startling new idea implicit in the definition of relativistic energy: *An object at rest has relativistic energy simply by virtue of its mass,* and this *mass-energy* is a part of the total energy that is conserved in an interaction within an isolated system.

9.7 THE SPATIAL PART OF FOUR-MOMENTUM

We have seen [Eq. (9.11)] that the spatial components P_x, P_y, P_z of an object's four-momentum vector become approximately equal to the components of the object's newtonian momentum at low velocities. Just as we defined an object's relativistic energy E to be the *time* component of its four-momentum, so we define an object's *relativistic momentum* p to be the magnitude of the spatial components of its four-momentum:

$$p \equiv \sqrt{P_x^2 + P_y^2 + P_z^2} \tag{9.27}$$

The relativistic momentum is thus the relativistic generalization of the *magnitude* of an object's newtonian momentum vector.

With the help of Eqs. (9.10), we can express an object's relativistic momentum in terms of its mass m and the speed v as follows:

$$p = \sqrt{\left(\frac{mv_x}{\sqrt{1-v^2}}\right)^2 + \left(\frac{mv_y}{\sqrt{1-v^2}}\right)^2 + \left(\frac{mv_z}{\sqrt{1-v^2}}\right)^2} = \frac{m\sqrt{v_x^2 + v_y^2 + v_z^2}}{\sqrt{1-v^2}}$$
$$= \frac{mv}{\sqrt{1-v^2}} \tag{9.28}$$

Note that p becomes approximately equal to the magnitude mv of the object's newtonian momentum when v is very much smaller than 1.

Equation (9.15), which expresses how an object's frame-independent mass m can be computed using the frame-dependent components of its four-momentum, can be written in terms of E and m as

$$m^2 = P_t^2 - (P_x^2 + P_y^2 + P_z^2) = E^2 - p^2 \tag{9.29}$$

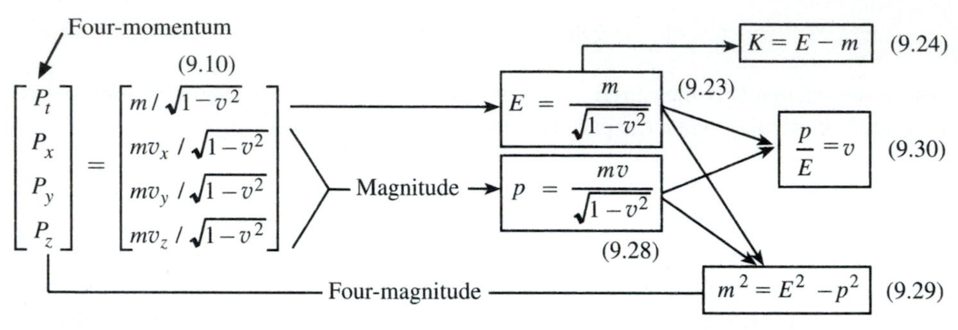

FIGURE 9.5
(Virtually) everything you need to know about four-momentum.

We can also use an object's relativistic energy E and relativistic momentum p in a given frame to determine the object's speed in that frame:

$$\frac{p}{E} = \frac{mv/\sqrt{1-v^2}}{m/\sqrt{1-v^2}} = v \tag{9.30}$$

This relationship applies to each individual spatial component as well:

$$\frac{P_x}{E} = \frac{mv_x}{\sqrt{1-v^2}}\,\frac{\sqrt{1-v^2}}{m} = v_x \qquad \text{similarly} \qquad \frac{P_y}{E} = v_y \qquad \frac{P_z}{E} = v_z \tag{9.31}$$

Indeed, the spatial components of the four-momentum can be thought of as expressing the rate at which relativistic energy is transported through space:

$$\begin{bmatrix} P_t \\ P_x \\ P_y \\ P_z \end{bmatrix} = \begin{bmatrix} m/\sqrt{1-v^2} \\ mv_x/\sqrt{1-v^2} \\ mv_y/\sqrt{1-v^2} \\ mv_z/\sqrt{1-v^2} \end{bmatrix} = \begin{bmatrix} E \\ Ev_x \\ Ev_y \\ Ev_z \end{bmatrix} \tag{9.32}$$

Equations (9.23), (9.24), and (9.28) through (9.30) express relationships between the quantities E, p, m, K, and v that are really helpful to know when working with the four-momentum. These core equations and the ways they connect to the four-momentum vector are summarized in Fig. 9.5.

9.8 AN EXAMPLE

Problem An alien spaceship (with mass m = 12,000 kg) is traveling relative to the earth with an x velocity of $v_x = -0.80$. What is the spaceship's relativistic energy E and relativistic momentum p? How is p related to the x component of the spaceship's four-momentum? What is its relativistic kinetic energy (express in kilograms and joules)?

Solution The speed of the spaceship is $v = +0.80 = 4/5$. According to the definitions of E and p [Eqs. (9.23) and (9.28), respectively],

$$E = \frac{m}{\sqrt{1-v^2}} = \frac{m}{\sqrt{1-16/25}} = \frac{m}{\sqrt{9/25}} = \frac{m}{3/5} = \frac{5m}{3} = 20{,}000 \text{ kg} \tag{9.33a}$$

$$p = \frac{mv}{\sqrt{1-v^2}} = \frac{m(4/5)}{3/5} = \frac{4m}{3} = 16{,}000 \text{ kg} \tag{9.33b}$$

Since the spaceship is moving in the $-x$ direction, we have $P_x = -p$ and $P_y = P_z = 0$, so the spaceship's four-momentum vector looks like

$$\begin{bmatrix} P_t \\ P_x \\ P_y \\ P_z \end{bmatrix} = \begin{bmatrix} E \\ -p \\ 0 \\ 0 \end{bmatrix} = \begin{bmatrix} 20{,}000 \text{ kg} \\ -16{,}000 \text{ kg} \\ 0 \\ 0 \end{bmatrix} \tag{9.34}$$

According to the definition of relativistic kinetic energy,

$$K = E - m = 20{,}000 \text{ kg} - 12{,}000 \text{ kg} = 8000 \text{ kg} \tag{9.35a}$$

$$K = 8000 \text{ kg} \left(\frac{3.0 \times 10^8 \text{ m}}{1 \text{ s}}\right)^2 \left(\frac{1 \text{ J}}{1 \text{ kg·m}^2/\text{s}^2}\right) = 7.2 \times 10^{20} \text{ J} \tag{9.35b}$$

This is the energy that has to be given to the spaceship to accelerate it from rest to $v = 4/5$. This energy is roughly the same as that released by 200,000 large (1-megaton) nuclear bombs or the yearly output of 20,000 large electrical power plants. Can you see that relativistic space travel is going to be an expensive proposition?

Note also that the spaceship's relativistic kinetic energy is substantially larger than would be predicted by the newtonian kinetic energy formula:

$$\tfrac{1}{2} mv^2 = (6000 \text{ kg})(4/5)^2 = 3840 \text{ kg} < K \tag{9.36}$$

This illustrates how at speeds v approaching that of light, $\frac{1}{2}mv^2$ becomes a poor approximation for the relativistic kinetic energy K.

PROBLEMS

9.1 TRANSFORMATION OF THE FOUR-MOMENTUM. Use the Lorentz transformation equations and the definition of the components of **P** to verify Eq. (9.12*b*).

9.2 THE MAGNITUDE OF THE FOUR-MOMENTUM. Verify Eq. (9.15) in each of the following ways.

a Use the facts that $\mathbf{P} = m\, d\mathbf{R}/d\tau$ and $d\tau^2 = dt^2 - dx^2 - dy^2 - dz^2$ for infinitesimally separated events.

b Square the definitions of the four-momentum components given in Eq. (9.10) and combine as required.

9.3 CONSERVATION OF FOUR-MOMENTUM. Show that if the total four-momentum of a two-object system is conserved in the Home Frame, then the time component of the four-momentum will be conserved in any Other inertial reference frame. [*Hint:* Equations (9.20) show that this is true for the x component of the total four-momentum.]

9.4 HIGH-VELOCITY LIMIT. Prove that in the limit as $v \to 1$, $E \approx p$. What velocity is required for these quantities to be equal to within 1 percent?

9.5 FOUR-MOMENTUM TO VELOCITY AND MASS. An object is observed in a certain frame to have a four-momentum of $[P_t, P_x, P_y, P_z] = [5.0 \text{ kg}, 3.0 \text{ kg}, 0, 0]$.

a Find the x velocity of the object in this frame.
b Find the object's mass.
c Find the object's kinetic energy.

9.6 FOUR-MOMENTUM TO VELOCITY AND MASS. An object in a certain frame is observed to have a four-momentum of $[P_t, P_x, P_y, P_z] = [13 \text{ kg}, -12 \text{ kg}, 0, 0]$.

a Find the object's x velocity in this frame.
b Find the object's mass.
c Find the object's kinetic energy.

9.7 FOUR-MOMENTUM TO VELOCITY AND MASS. An object is observed in a certain frame to have a four-momentum of $[P_t, P_x, P_y, P_z] = [18 \text{ kg}, 9.0 \text{ kg}, 15 \text{ kg}, 1.0 \text{ kg}]$.

a Find the object's velocity (vector).
b Find the object's speed.
c Find the object's mass.
d Find the object's relativistic momentum p.
e Find the object's kinetic energy.

9.8 VELOCITY AND MASS TO FOUR-MOMENTUM. Imagine that an object of mass 5.0 kg has velocity components $v_x = -0.866$, $v_y = v_z = 0$.

a Find the object's total energy E.
b Find the object's relativistic momentum p.
c Find the spatial components P_x, P_y, and P_z of the object's four-momentum.
d Find the object's relativistic kinetic energy K.

9.9 VELOCITY AND MASS TO FOUR-MOMENTUM. Imagine that a dust particle of mass 2.0 μg is traveling at a speed of 4/5 in the xy plane at an angle of 30° clockwise from the x axis.

a Find the particle's total energy E.
b Find the particle's relativistic momentum p.
c Find the particle's spatial four-momentum components P_x, P_y, and P_z.
d Find the particle's kinetic energy K (in joules).

9.10 LETTER BOMB. To compete with E-Mail, the Post Office offers a new service called Super Express Mail, where a letter is sent to its destination at a speed of 0.999 using a special letter accelerator. If the letter has a mass of 25 g (typical for a letter), compute its relativistic kinetic energy at its cruising speed. Electrical energy costs about $\$0.03/10^6$ J. If the letter accelerator is 100 percent efficient, what will be the approximate cost of the letter's stamp? If the letter misses its target and hits a nearby building, describe the consequences. (*Hint:* A typical nuclear bomb releases about 4×10^{14} J when it explodes.)

9.11 YOU ARE WORTH A LOT! If electrical energy can be sold at $\$0.03/10^6$ J, compute how much your rest energy is worth in dollars. That is, find the amount of money that your survivors could put in your memorial fund if there was a way to convert your mass entirely into electrical energy. (It is probably a good thing that this is not easy to do.)

10

CONSERVATION OF FOUR-MOMENTUM

God is cunning, but not malicious.

Einstein[1]

10.1 OVERVIEW

In the last chapter, we saw that conservation of the total ordinary (newtonian) momentum of an isolated system of objects is *inconsistent* with the principle of relativity, because of the complicated nature of the Einstein velocity transformation. On the other hand, we saw that conservation of the total four-momentum of an isolated system *is* consistent with the principle of relativity. Moreover, the four-momentum of an object reduces to the newtonian momentum in the low-velocity limit. Therefore, if anything like "momentum" is to be conserved, it must be in fact four-momentum that is conserved. Let us assume that this is so.

The purpose of this chapter is to explore some of the surprising consequences and experimental tests of this assumption. First of all, we will discuss how to draw energy-momentum diagrams of collision processes and what such diagrams can tell us. Then, using such diagrams, we will explore in some depth the assertion (hinted at in the last chapter) that mass is just another form of (relativistic) energy, and processes do exist that can convert mass to energy, and vice versa. We will also learn how light itself can be represented by a four-momentum vector. Finally, we will apply the law of

[1]Quoted in A. P. French (ed.), *Einstein: A Centenary Volume,* Cambridge, Mass.: Harvard, 1979, p. 73.

conservation of four-momentum to a variety of examples, ranging from practical experimental tests to speculations about relativistic space travel.

10.2 ENERGY-MOMENTUM DIAGRAMS

The four-momentum of an object moving in the spatial x direction can be visually represented as an arrow on a special kind of spacetime diagram called an *energy-momentum* diagram (see Fig. 10.1). Just as the direction of the arrow representing an object's ordinary momentum is tangent to its path through space, the direction of the arrow representing an object's four-momentum is tangent to its worldline in spacetime (because the object's four-momentum vector $\mathbf{P} = m\, d\mathbf{R}/d\tau$ at any given point along its worldline is proportional to the object's differential displacement in spacetime $d\mathbf{R}$ along that worldline around that point) (see Fig. 10.2).

Since the inverse slope of an object's worldline at any instant is equal to its x velocity at that instant, the inverse slope of the object's four-momentum arrow at a given time (i.e., run/rise = P_x/E) should also be equal to its x velocity at that time if the two vectors are to be parallel. Equation (9.31) in the previous chapter says essentially the same thing:

$$\frac{P_x}{E} = \frac{mv_x}{\sqrt{1-v^2}} \frac{\sqrt{1-v^2}}{m} = v_x \tag{10.1}$$

The *four-magnitude* of an object's four-momentum is its mass m [see Eq. (9.15)]: this value is frame-independent and independent of the object's motion in a given frame. But the *length* of the arrow representing the object's four-momentum on an energy-momentum diagram depends on the object's velocity: the length of the arrow on the diagram is *not* proportional to the four-magnitude of the corresponding four-momentum. (This is analogous to the problem with the spacetime interval discussed in Sec. 4.5.) In fact, according to Eq. (9.29), we have (assuming that the object is moving in the x direction so that $P_y = P_z = 0$ and $p = |P_x|$)

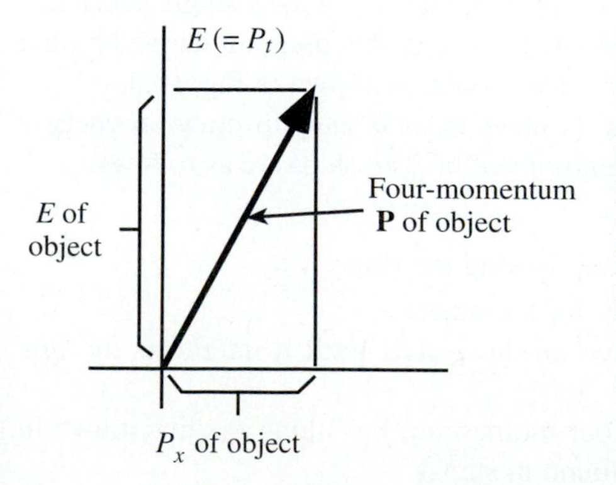

FIGURE 10.1
Energy-momentum diagram showing the four-momentum of a certain object moving in the $+x$ direction. The object's four-momentum is represented on the diagram by an arrow. The projections of this arrow on the vertical and horizontal axes represent the values of the object's relativistic energy E (= P_t) and its relativistic x momentum P_x. Note that the object's relativistic momentum $p = |P_x|$.

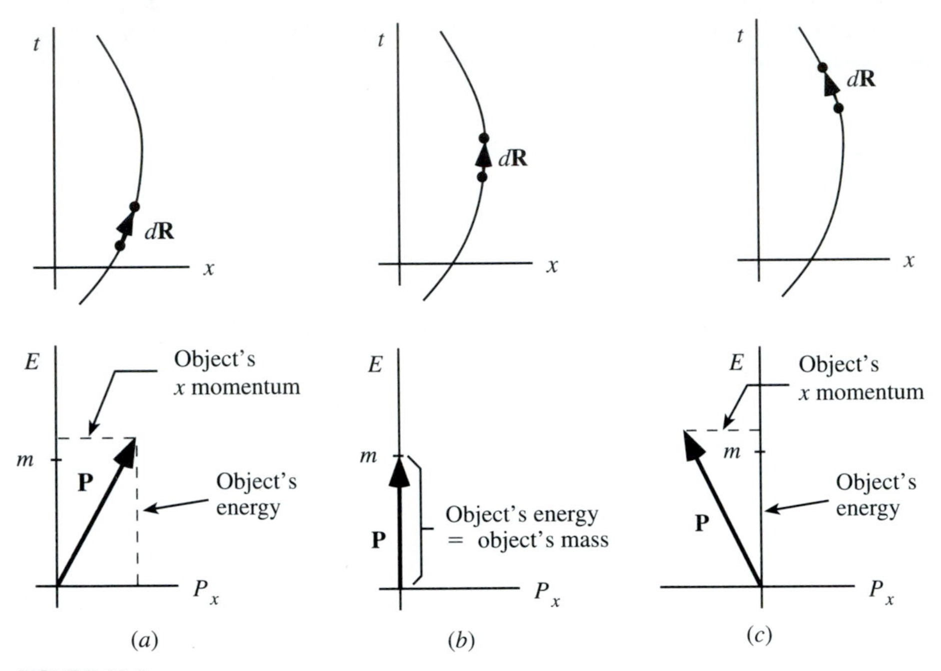

FIGURE 10.2
(*a*) At any given time, the arrow representing an object's four-momentum on an energy-momentum diagram points in a direction tangent to the object's worldline, because **P** is proportional to *d***R**. (*b*) When an object is at rest (even if just at an instant), its four-momentum is vertical and its energy is equal to its mass [see Eq. (9.10*a*)]. (*c*) When the object moves in the $-x$ direction, its *x* momentum is negative [see Eq. (9.10*b*)] but its energy remains positive (and indeed greater than its mass).

$$m^2 = E^2 - p^2 = E^2 - (P_x)^2 \tag{10.2}$$

This means that the tips of the four-momentum arrows for objects of identical mass m traveling at different x velocities (or the four-momentum arrows for a single accelerating object observed at different times) lie along a curve on the diagram defined by the equation $m^2 = E^2 - p^2$. This curve is in fact a hyperbola, as shown in Fig. 10.3.

If you know an object's x velocity and its mass m, it is easy to draw an energy-momentum diagram showing its four-momentum vector. The steps are as follows:

1 Set up your E and P_x axes.
2 Draw a line from the origin of those axes having the slope $1/v_x$.
3 Compute the value of $E = m/\sqrt{1 - v_x^2}$ for the object.
4 Draw a horizontal line from this value on the E axis until it intercepts the line that you drew in step 2.
5 The arrow representing the object's four-momentum lies along the line drawn in step 2, with its tip at the intersection found in step 4.

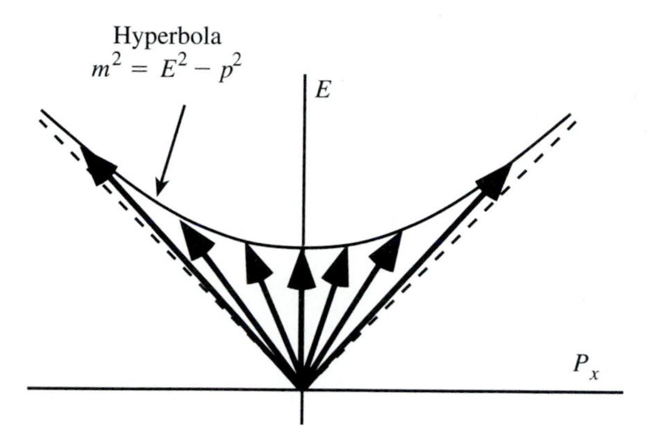

FIGURE 10.3
Energy-momentum diagram showing the four-momentum arrows for a set of identical objects of mass m moving at different x velocities in the Home Frame. The tips of all these arrows lie on the hyperbola defined by the equation $m^2 = E^2 - p^2$. Note that as the object's x velocity approaches ±1 (and thus p/E approaches 1), both p and E have to become very large if the difference of their squares is to remain fixed.

We can easily read an object's relativistic kinetic energy $K = E - m$ directly from an energy-momentum diagram. For example, K for an object of mass m moving at a speed $v = 3/5$ is $m/4$, as shown in Fig. 10.4. (Note that $m/4 \neq \frac{1}{2}m(3/5)^2$ but in fact is substantially larger!) Figures 10.3 and 10.4 together make it clear that as an object's speed v approaches 1 (the speed of light), both the object's total relativistic energy E and its kinetic energy K go to infinity. This means that *you would have to supply an infinite amount of energy to accelerate an object of nonzero mass to the speed of light.* (This is the most practical reason that no object can go faster than the speed of light: all the energy in the universe could not accelerate even a mote of dust to that speed!)

Virtually all that you need to know to construct and interpret an energy-momentum diagram is summarized in Fig. 10.5.

10.3 CONSERVATION OF FOUR-MOMENTUM: AN EXAMPLE

The law of conservation of four-momentum (like the law of conservation of ordinary momentum) is most useful when applied to an isolated system of objects undergoing

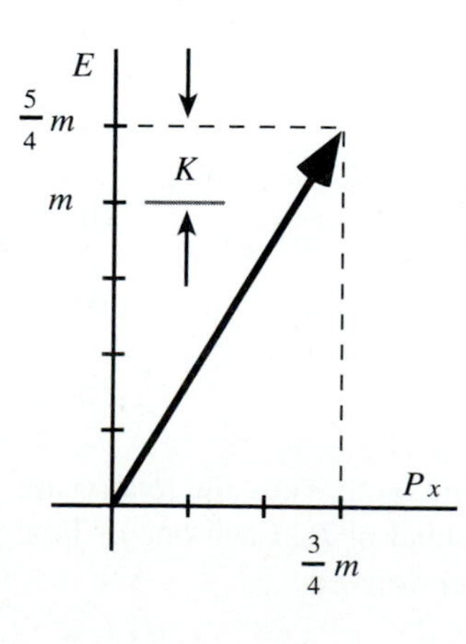

FIGURE 10.4
An energy-momentum diagram of an object of mass m traveling at an x velocity $v_x = 3/5$. The four-momentum arrow for such an object has a slope of 5/3 and an energy of $5m/4$, since $\sqrt{1-(3/5)^2} = \sqrt{1-9/25} = \sqrt{16/25} = 4/5$, implying that $E = m/\sqrt{1-v_x^2} = 5m/4$. For the arrow to have the correct slope, we must have $P_x = 3m/4$. Note that the object's relativistic kinetic energy $K = E - m = m/4$ can be read directly from the diagram.

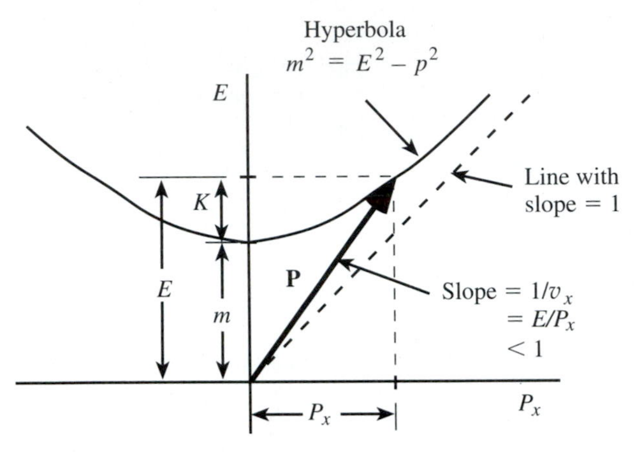

FIGURE 10.5
Virtually everything that you need to know about four-momentum diagrams. No matter what the x velocity of an object of mass m might be, the tip of its four-momentum arrow lies on the hyperbola $m^2 = E^2 - p^2$ for whatever that object's mass m might be. The inverse slope of the four-momentum arrow is equal to v_x, which always has a magnitude less than 1.

some type of *collision* process (i.e., some kind of sudden interaction between the objects in the system that may be strong and complicated but limited in time). In such a case, the system has a clearly defined state "before" and "after" the collision, making it easy to compute the total four-momentum in the system both before and after the collision. The law of conservation of four-momentum states that the system should have the *same* total four-momentum after the collision process as it had before.

What does this really mean mathematically? Since four-momentum is a (four-dimensional) vector quantity, conservation of four-momentum means that *each component of the system's total four-momentum is separately conserved.* For example, consider a system consisting of two objects, and let the objects' four-momenta before the collision be $\mathbf{P}_1$ and $\mathbf{P}_2$, and after the collision be $\mathbf{P}_3$ and $\mathbf{P}_4$. Conservation of four-momentum requires that

$$E_1 + E_2 = E_3 + E_4 \tag{10.3a}$$

$$P_{1x} + P_{2x} = P_{3x} + P_{4x} \tag{10.3b}$$

$$P_{1y} + P_{2y} = P_{3y} + P_{4y} \tag{10.3c}$$

$$P_{1z} + P_{2z} = P_{3z} + P_{4z} \tag{10.3d}$$

remembering that the time component of a four-momentum vector (i.e., the relativistic energy) is usually given the more evocative symbol E instead of P_t. Each one of Eqs. (10.3) has to be *separately* true for four-momentum to be conserved.

In this chapter, we will focus primarily on objects moving in only *one* dimension, which we can take to be the $\pm x$ direction. This simplifies the mathematics significantly without any substantial loss of understanding, allowing us to ignore the y and z components of the four-momenta (which are always zero) and focus on the t and x components [i.e., Eqs. (10.3*a*) and (10.3*b*)].

Let us consider a specific example. Imagine that somewhere in deep space a certain rock with mass $m_1 = 12$ kg is moving in the $+x$ direction with $v_{1x} = +4/5$ in some inertial reference frame. This rock then strikes another rock of mass $m_2 = 28$ kg at rest ($v_{2x} = 0$). Instead of instantly vaporizing into a cloud of gas (as any real rocks colliding at this speed would), let us pretend that the first rock simply bounces off the more massive rock and is subsequently observed to have an x velocity $v_{3x} = -5/13$. What is the x velocity v_{4x} of the larger rock after the collision?

The first step in solving this problem is to calculate the energy E_1 and the x momentum P_{1x} of the smaller rock before the collision. Using the definitions of these four-momentum components, we find that

$$E_1 \equiv \frac{m_1}{\sqrt{1 - v_{1x}^2}} = \frac{m_1}{\sqrt{1 - (4/5)^2}} = \frac{m_1}{\sqrt{9/25}} = \frac{m_1}{3/5} = \frac{5(12 \text{ kg})}{3} = 20 \text{ kg} \tag{10.4a}$$

$$P_{1x} \equiv \frac{m_1 v_{1x}}{\sqrt{1 - v_{1x}^2}} = \frac{m_1(+4/5)}{3/5} = \frac{4(12 \text{ kg})}{3} = +16 \text{ kg} \tag{10.4b}$$

Similarly, the energy and x momentum of the larger rock before the collision are

$$E_2 \equiv \frac{m_2}{\sqrt{1 - v_{2x}^2}} = \frac{m_2}{\sqrt{1 - (0)^2}} = m_2 = 28 \text{ kg} \tag{10.5a}$$

$$P_{2x} \equiv \frac{m_2 v_{2x}}{\sqrt{1 - v_{2x}^2}} = \frac{m_1(0)}{\sqrt{1 - 0^2}} = 0 \text{ kg} \tag{10.5b}$$

The energy and x momentum of the smaller rock *after* the collision are

$$E_3 \equiv \frac{m_1}{\sqrt{1 - v_{3x}^2}} = \frac{m_1}{\sqrt{1 - (-5/13)^2}} = \frac{m_1}{\sqrt{144/169}} = \frac{m_1}{12/13} = \frac{13(12 \text{ kg})}{12} = 13 \text{ kg} \tag{10.6a}$$

$$P_{3x} \equiv \frac{m_1 v_{3x}}{\sqrt{1 - v_{3x}^2}} = \frac{m_1(-5/13)}{12/13} = \frac{-5(12 \text{ kg})}{12} = -5 \text{ kg} \tag{10.6b}$$

Conservation of four-momentum requires that the four-momentum vectors before the collision add up to the same value after the collision:

$$\begin{matrix} t \text{ component:} \\ x \text{ component:} \end{matrix} \quad \begin{bmatrix} E_1 \\ P_{1x} \end{bmatrix} + \begin{bmatrix} E_2 \\ P_{2x} \end{bmatrix} = \begin{bmatrix} E_3 \\ P_{3x} \end{bmatrix} + \begin{bmatrix} E_4 \\ P_{4x} \end{bmatrix} \tag{10.7a}$$

$$\begin{bmatrix} E_4 \\ P_{4x} \end{bmatrix} = \begin{bmatrix} E_1 \\ P_{1x} \end{bmatrix} + \begin{bmatrix} E_2 \\ P_{2x} \end{bmatrix} - \begin{bmatrix} E_3 \\ P_{3x} \end{bmatrix} = \begin{bmatrix} 20\text{ kg} \\ 16\text{ kg} \end{bmatrix} + \begin{bmatrix} 28\text{ kg} \\ 0 \end{bmatrix} - \begin{bmatrix} 13\text{ kg} \\ -5\text{ kg} \end{bmatrix} = \begin{bmatrix} 35\text{ kg} \\ +21\text{ kg} \end{bmatrix} \tag{10.7b}$$

Knowing the energy and momentum of an object is sufficient information to determine both its mass and velocity. Using Eq. (9.29), we see that the mass of the larger rock is

$$m = \sqrt{E_4^2 - P_{4x}^2} = \sqrt{(35\text{ kg})^2 - (21\text{ kg})^2} = (7\text{ kg})\sqrt{5^2 - 3^2} = (7\text{ kg})4 = 28\text{ kg} \tag{10.8a}$$

after the collision (just as it was before). According to Eq. (9.31), its final x velocity is

$$v_{4x} = \frac{P_{4x}}{E_4} = \frac{+21\text{ kg}}{35\text{ kg}} = \frac{3}{5} \tag{10.8b}$$

You can easily show in this case (see Prob. 10.3) that newtonian momentum is *not* conserved by the collision just described.

10.4 SOLVING CONSERVATION PROBLEMS GRAPHICALLY

It is possible to solve a conservation of four-momentum problem (like the example just considered) *graphically* using an energy-momentum diagram. The sum of four-momenta is defined like the sum of ordinary vectors (you simply add the components), so you can add four-momentum arrows on an energy-momentum diagram just as you would ordinary vector arrows (by putting the tail of one vector on the tip of the other while preserving their directions). Using this technique, we see in Fig. 10.6*a* that in the rock example, the system's total four-momentum *before* the collision has components $E_T = 48$ kg, $P_{T,x} = 16$ kg. The two rocks' four-momentum arrows after the collision have to add up to the *same* total four-momentum arrow, and since we know the smaller rock's four-momentum after the collision, we can *construct* the larger rock's final four-momentum arrow (Fig. 10.6*b*). We can then read the components of this arrow right off the diagram, getting the same results as in Eqs. (10.7).

This kind of graphical approach to the problem is not usually much easier than the algebraic approach, but it does have some advantages: (1) It provides a more visual and concrete way of dealing with the problem and may be helpful to you if you find the algebraic approach rather abstract. (2) When used in conjunction with the algebraic method, it serves as a useful check on the algebraic results: it is more difficult to make an error using the graphical method. (3) In some cases, as we will see, simply *looking* at the diagram can yield qualitative information that is very difficult to get from the algebraic equations alone.

In short, the graphical method represents an alternative method for solving problems involving conservation of four-momentum that often complements the algebraic approach. Armed with both these techniques, we are now ready to explore some of the strange and interesting consequences of the law of conservation of four-momentum.

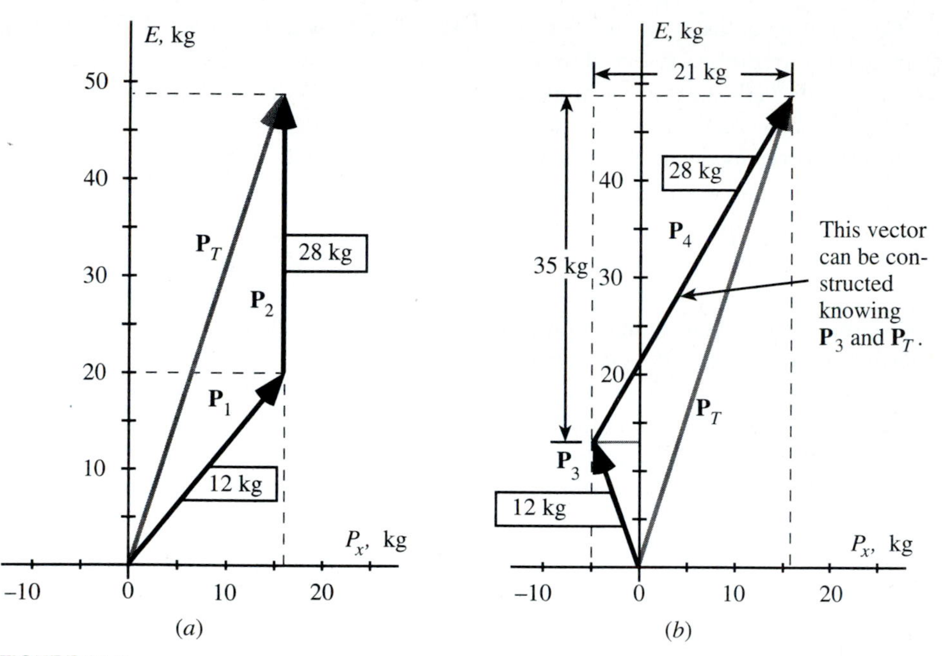

FIGURE 10.6
(a) The four-momenta of the rocks *before* the collision. The vector sum of these four-momenta is represented by the arrow $\mathbf{P}_T$. Since the magnitudes of the individual four-momenta arrows (which equal the masses of the corresponding rocks) cannot be read directly from the diagram, I have adopted the expedient of attaching a "flag" to each four-momentum arrow that states its magnitude. *(b)* The four-momenta of the rocks *after* the collision. The vector sum of these rocks' four-momenta is still $\mathbf{P}_T$ by four-momentum conservation. Since $\mathbf{P}_3$ is also known, it is possible to construct the unknown four-momentum $\mathbf{P}_4$, read its components from the diagram as shown, and compute its corresponding mass and *x* velocity.

10.5 THE MASS OF A SYSTEM OF PARTICLES

As we have seen, the relativistic energy of an object is not simply equal to its kinetic energy (even at low velocities) but involves the mass of the object as well. The fact that $E = P_t$ is conserved by the internal interactions of an isolated system does not imply that the mass of an object and its kinetic energy are separately conserved, only that the sum of these two things are conserved. This implies that processes that convert mass to kinetic energy, and vice versa, do not necessarily violate the law of conservation of four-momentum and therefore may exist. Mass and kinetic energy are seen in the theory of special relativity to be simply two parts of the same whole (the relativistic energy). There is no reason to presuppose a barrier between these two manifestations of relativistic energy that would preclude the conversion of one into the other.

In much of the remainder of this chapter, we will be considering a variety of examples of processes that do just that. We will begin with a simple example that illustrates a crucial thing we need to understand about "mass" before we can go further: *the mass of a system is generally different from the sum of the masses of its parts.*

Consider the collision of two identical balls of putty with mass $m = 4$ kg which in some inertial frame are observed to have x velocities of $v_{1x} = +3/5$ and $v_{2x} = -3/5$; that is, these putty balls are approaching each other with equal speeds. Imagine that when these putty balls collide, they stick together, as shown in Fig. 10.7. Note that before the collision, the x component of the system's total four-momentum is zero:

$$P_{1x} + P_{2x} = \frac{m(+3/5)}{\sqrt{1-(3/5)^2}} + \frac{m(-3/5)}{\sqrt{1-(-3/5)^2}} = \frac{m(3/5 - 3/5)}{\sqrt{1-9/25}} = 0 \tag{10.9}$$

so conservation of four-momentum implies that the x component of the final mass' four-momentum is zero as well, meaning it must be at rest.

What of relativistic energy conservation in this case? A newtonian analysis of this collision would speak of the kinetic energy being converted into thermal energy in this inelastic collision. Such an analysis would also assert that the mass of the coalesced particle is $M = m + m = 2m$. But we have more constraints to consider in a relativistic solution to this problem. If the spatial components of the four-momentum are conserved in this collision, the time component must also be conserved, whether the collision is elastic or not. But how can we think of the relativistic energy being conserved in this case, since no mention has been made of thermal energy in the definition of the relativistic energy given in Chap. 9?

The answer is direct and surprising. Since the final object is motionless, its relativistic energy is simply equal to its mass M. But by conservation of four-momentum, we have

$$M = E_1 + E_2 = \frac{m}{\sqrt{1-(3/5)^2}} + \frac{m}{\sqrt{1-(-3/5)^2}} = \frac{2m}{\sqrt{16/25}} = \frac{10m}{4} = 10 \text{ kg} \tag{10.10}$$

which is *not* equal to $2m = 8$ kg! Conservation of four-momentum thus requires that the final object have a *greater* mass than the sum of the masses that collided to form it!

We know from experience with collisions at low speeds that when two objects collide and stick together, their energy of motion gets converted to thermal energy: the final object is a little warmer than the original objects. (In this case, actually, the final object will be a *lot* warmer than the original objects, so much so that any *real* putty balls colliding at such speeds would vaporize instantly.) What Eq. (10.10) is telling us

FIGURE 10.7
The inelastic collision of two putty balls as seen in the frame where they initially have equal speeds but opposite directions. $m = 4.0$ kg.

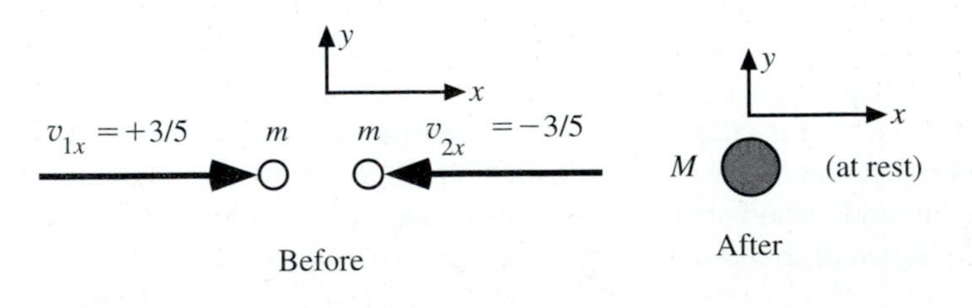

is that the final object *has* to be more massive than the original objects and that the increased thermal energy is somehow correlated with this.

But where does this extra mass actually reside? The final object has the same number of atoms as the original objects did. Does each atom gain some extra mass somehow? This seems absurd. The increased thermal energy in the final object means that its atoms will jostle around more vigorously. Can the motion of these atoms "have mass" in some sense? This seems crazy: *individual* particles have the same mass no matter how they move. *So where is this extra mass?*

There is only one fully self-consistent way to answer this question: *the extra mass is a property of the system as a whole* and does not reside in any of its parts.

This can be vividly illustrated as follows. Consider the "system" consisting of the two balls of putty *before* they collide. If they are considered separate objects, the putty balls each have a mass m of 4 kg and a relativistic energy of 5 kg and one has an x momentum of −3 kg and the other +3 kg. On the other hand, if we consider the balls to constitute a *system,* the system has a total x momentum of zero and a total energy of 10 kg, meaning that its mass $M \equiv \sqrt{E_T^2 - p_T^2}$ is equal to 10 kg. (This is illustrated by Fig. 10.8.) So we see that the thermal energy produced by the collision is *not* the source of this extra mass: the extra mass was present in the "system" *before* the collision and remains the same after the collision.

FIGURE 10.8
(a) Putty balls before the collision, considered as two individual objects. Each object has its own mass, energy, and x momentum. *(b)* Putty balls before the collision, considered a single system. The system's x momentum is zero, meaning that its total energy of 10 kg is also equal to its mass.

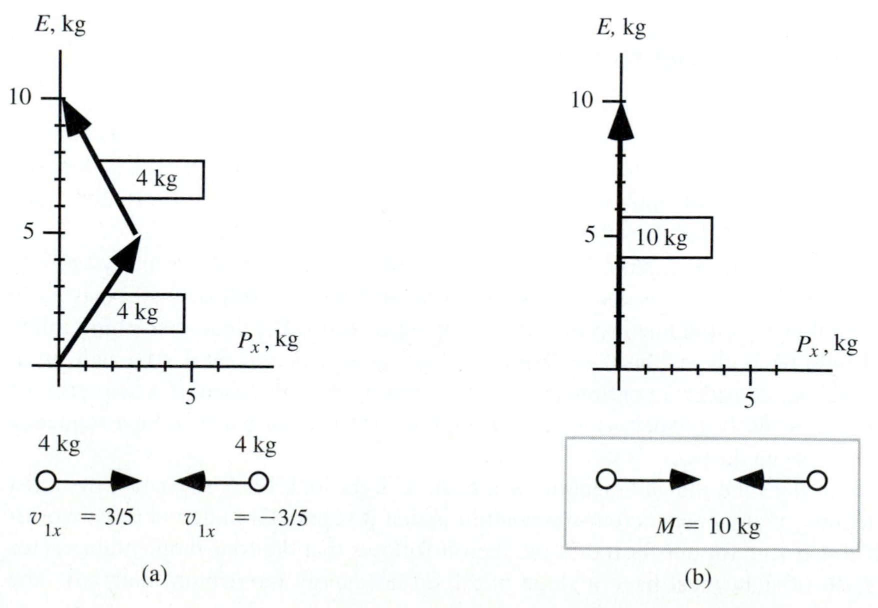

So in some sense, mass is not "created" by the collision process at all: the collision simply *manifests* the mass of the *system* of two initial objects in the mass of a *single* final object. If we focus on the masses of the individual objects in the system before and after the collision, we think of mass being created. But if we focus on the *system* before and after the collision, we see that its mass remains the same. It is possible to get unnecessarily hung up on the difference between the mass of a system and the masses of its parts. The reason this seems screwy is that we are *used* to treating mass as if it were additive: the mass of a jar of beans is the sum of the masses of the individual beans plus the mass of the container, right? This is true enough at low velocities. But if we had common experience with a jar full of beans that bang around inside the jar with speeds close to that of light, then we would be *used* to the idea that such a jar would have a different mass than the mass of the individual beans. Mass is simply *not* additive in the way that energy and x momentum are.

There are actually many examples of things in the world where the whole is greater than the sum of its parts. The meaning of a poem is not the same as the sum of the meaning of the individual letters in the words. The "life" of an organism cannot be localized in any of its parts. We simply need to start thinking about mass in the same way as we think about these things.

The best way to look at this is to think of the mass of a system of particles as *a property of the system as a whole* (i.e., the magnitude of the system's total four-momentum vector) and something that simply does not have very much to do with the masses of its parts. The only self-consistent way to define the mass of a *system* is as the magnitude of the system's total four-momentum, and if this definition leads to the mass of a system being greater or less than the masses of its parts, well, that's the way it is!

10.6 THE FOUR-MOMENTUM OF LIGHT

We all have experienced the fact that light carries energy: we have felt sunlight warm our skin or seen an electric motor powered by solar cells or learned that plants convert the energy in sunlight into chemical energy. Since we have seen that energy is the time component of four-momentum, it follows that light should have an associated four-momentum vector. What does the four-momentum of light look like?

Previously, we have explored the four-momenta of objects (rocks or putty balls or the like) that could be considered to be *particles* that have a well-defined position in space and thus a well-defined worldline through spacetime. The analogous thing in the case of light would be a "flash" or "burst" of light energy that is similarly localized in space. We can consider a continuous beam of light to be composed of a sequence of closely spaced flashes, much as we might imagine a stream of water to be a sequence of closely spaced drops.

So what does the four-momentum of a flash of light look like? Arguably, the most basic feature of any object's four-momentum is that it is parallel to that object's worldline. If this is true for our flash of light, then it follows that the four-momentum vector for a flash of light must have a slope of ± 1 on an energy-momentum diagram. The

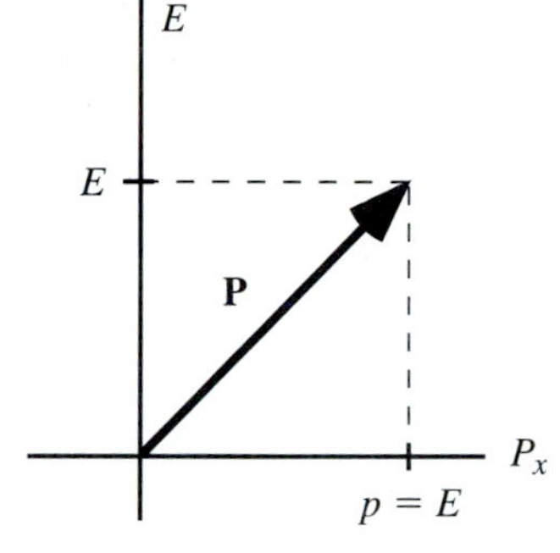

FIGURE 10.9
An energy-momentum diagram showing the four-momentum of a flash of light moving in the $+x$ direction. If the four-momentum is to be parallel to the flash's worldline, then it must be drawn with slope +1 on the energy-momentum diagram. This implies that the flash's relativistic momentum p must have the same value (in SR units) as its energy: $p = E$.

four-momentum vector of a flash with a given energy E moving in the $+x$ direction will thus look something like the vector shown in Fig. 10.9.

You can see from this diagram that if the flash's four-momentum vector is to have such a slope, it must have a spatial relativistic momentum p equal to its relativistic energy E. This is in fact consistent with Eq. (9.30), which in the case of light tells us that

$$\frac{p}{E} = v = 1 \quad \Rightarrow \quad p = E \qquad \text{(for a light flash)} \tag{10.11}$$

One immediate implication of this important formula is that *light must carry momentum* (as well as energy). Light bouncing off a mirror will thus transfer momentum to the mirror (causing it to recoil) in much the same way that a ball bouncing off an object transfers momentum to the object and causes it to recoil. This has been experimentally verified,[1] and it is now known that the pressure exerted by light due to its momentum plays an important part in the evolution of stars, the evolution of the early universe, and many other astrophysical processes.

Another immediate consequence is that a flash of light has zero mass. We have defined the mass of an object in special relativity to be the invariant magnitude of its four-momentum. According to Eq. (10.11), the mass of a flash of light is

$$m^2 = E^2 - p^2 = 0 \tag{10.12}$$

This is actually a good thing. If a flash of light were to have some nonzero mass m, then its energy $E = m/\sqrt{1 - v^2}$ and relativistic momentum $p = mv/\sqrt{1 - v^2}$ would both have to be infinite, since $v = 1$ for light, which makes the denominator zero in each expression. But since $m = 0$ as well, these equations instead read $E = 0/0$ and $p = 0/0$: the ratio 0/0 is "undefined" instead of being infinite, meaning that the equations

[1]Maxwell's classical theory of electromagnetic waves also predicts that light should carry momentum $p = E/c$; this was understood well before special relativity. Experiments performed in 1903 by Nichols and Hull in the United States and Lebedev in Russia confirmed this experimental prediction. See G.E. Henry, "Radiation Pressure," *Sci. Am.*, June 1957, p. 99.

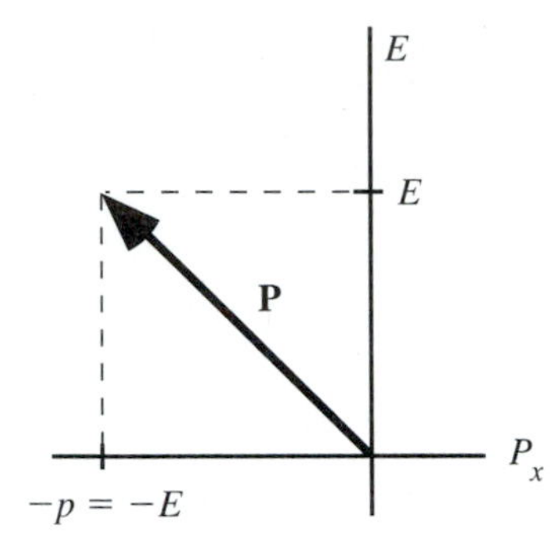

FIGURE 10.10
An energy-momentum diagram showing the four-momentum of a flash of light moving in the $-x$ direction. In this case, its x momentum is negative: $P_x = -p$, where p is equal to the *magnitude* of the flash's spatial momentum ($p = |P_x|$ in this case). Note that we still have $p = E$: this is a general relation for light, independent of the direction the flash is traveling.

$E = m/\sqrt{1 - v^2}$ and $p = mv/\sqrt{1 - v^2}$ simply do not tell us anything useful about the four-momentum of light.

If a light flash is moving in the $-x$ direction instead of the $+x$ direction, the slope of its four-momentum arrow on an energy-momentum diagram is -1 instead of $+1$ and its x momentum is negative: $P_x = -p = -E$ (note that p and E are positive by definition), as shown in Fig. 10.10.

10.7 EXAMPLE: A MATTER-ANTIMATTER ROCKET

When antimatter is brought into contact with matter, they annihilate each other, converting their mass-energy entirely into light. A perfect rocket engine might mix antimatter with an equal amount of matter and direct the resulting light in a tight beam out of the "nozzle" of the engine: no other kind of "exhaust" could possibly carry more momentum out the rear of the rocket per unit energy expended than light can. Imagine a rocket of original mass $M = 90{,}000$ kg sitting at rest in some frame in deep space (M includes the mass of the matter-antimatter fuel). Imagine that it fires its engines, emitting a burst of light having a total (unknown) energy E_L. If after doing this the ship has a final speed of $v = 4/5$, what is its final mass m (see Fig. 10.11)?

The law of conservation of four-momentum tells us that the total four-momentum of an isolated system of interacting objects will be conserved. The initial four-momentum of the system in this case is the four-momentum of a mass M at rest, which has components $P_t = M$ and $P_x = P_y = P_z = 0$. After the engines have fired, the

FIGURE 10.11
The hypothetical situation being described in this section.

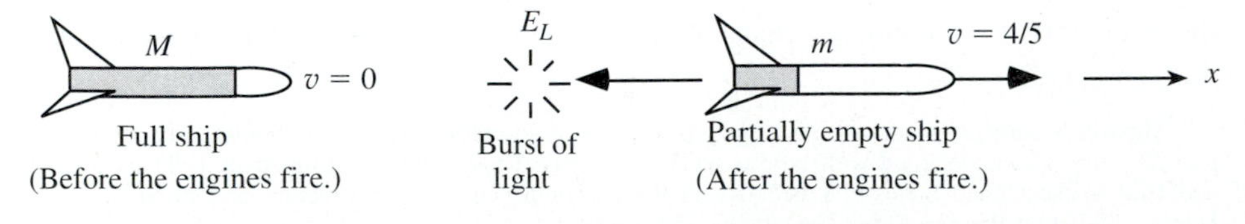

system consists of a light flash and the somewhat lighter ship. Let us define the $+x$ direction to be the ship's final direction of motion, as shown in the diagram. According to Eq. (10.11) the flash's relativistic momentum p_L will be equal to its energy E_L; since the flash is moving in the $-x$ direction, we will have $P_x = -p_L = E_L$ and $P_y = P_z = 0$. We do not know the final ship mass m, but we do know that it is moving at a speed $v = 4/5$ in the $+x$ direction, so by Eqs. (9.10) we know that the final ship's four-momentum has components $P_t = m/\sqrt{1-v^2} = m/\sqrt{1-(4/5)^2} = m/\sqrt{9/25} = 5m/3$, $P_x = +p = Ev = 4m/3$, and $P_y = P_z = 0$. Putting this all together, then, the law of conservation of four-momentum reads

$$\begin{bmatrix} M \\ 0 \\ 0 \\ 0 \end{bmatrix} = \begin{bmatrix} E_L \\ -E_L \\ 0 \\ 0 \end{bmatrix} + \begin{bmatrix} \frac{5}{3}m \\ +\frac{4}{3}m \\ 0 \\ 0 \end{bmatrix} \tag{10.13}$$

The two component equations on the bottom of this vector equation simply say that $0 = 0 + 0$, which we knew already. The top two equations tell us that

$$M = E_L + \frac{5}{3}\, m \tag{10.14a}$$

$$0 = -E_L + \frac{4}{3}\, m \quad \Rightarrow \quad E_L = \frac{4}{3}\, m \tag{10.14b}$$

These equations represent two equations in the two unknowns E_L and m. We can solve these equations for m by plugging (10.14b) into (10.14a) and eliminating E_L:

$$M = \frac{4}{3}m + \frac{5}{3}m = \frac{9}{3}m = 3m \quad \Rightarrow \quad m = \frac{1}{3}M = 30{,}000 \text{ kg} \tag{10.15}$$

So even this perfect rocket has to use 60,000 kg of fuel to boost the remaining 30,000 kg to a speed of 0.8.

Figure 10.12 shows an energy-momentum diagram of the situation. The ship's original four-momentum is represented by a vertical arrow with vertical length M. Since we know that the flash of light is moving in the $-x$ direction with speed 1, its four-momentum will have a slope of -1. Similarly, since we know that the final ship's speed is 4/5 in the $+x$ direction, its four-momentum arrow will have a slope of 4/5 on the diagram. These arrows have to add up to the original ship's four-momentum, so if we construct a line with slope -1 down from the tip of this arrow and another line with slope 4/5 up from the base, the point where these lines intersect determines the base and tip of the light and final ship four-momentum vectors, respectively, as shown. We can read from this that the ship's final four-momentum has components $E = 50{,}000$ kg and $P_x = 40{,}000$ kg, implying that its mass is $m = \sqrt{E^2 - p^2} = \sqrt{(50{,}000 \text{ kg})^2 - (40{,}000 \text{ kg})^2} = 30{,}000$ kg, as found above.

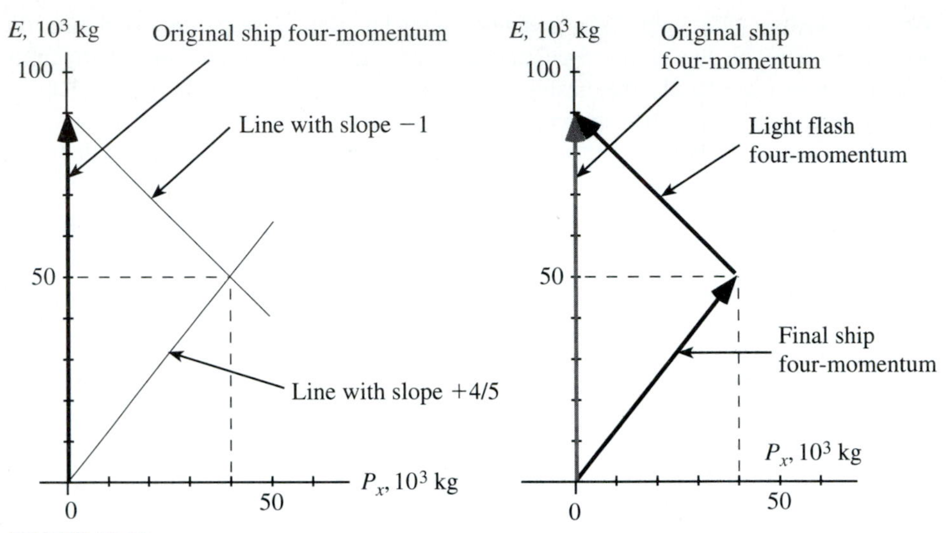

FIGURE 10.12
(a) To solve the problem graphically, first construct the ship's original four-momentum, and then draw a line with slope −1 and a line with slope 4/5. The flash four-momentum and final ship four-momentum have to be parallel to these lines, respectively. *(b)* Draw the four-momentum vectors for the flash and final ship parallel to these lines so that they add up to the original ship's four-momentum. From this construction, we can read the components of the final ship's four-momentum and compute its mass.

PROBLEMS

10.1 AN ELASTIC COLLISION. An object with mass $m_1 = 8$ kg traveling with an x velocity of $v_{1x} = 15/17$ collides with an object with mass $m_2 = 12$ kg traveling with an x velocity of $v_{2x} = -5/13$. After the collision, the 8-kg object is measured to have an x velocity of $v_{3x} = -3/5$.

a Find the relativistic energy and momentum of the other object.

b Find its x velocity.

c Show that it has the same mass as it started with.

Solve this problem both algebraically and graphically.

10.2 AN INELASTIC COLLISION. A particle of mass m traveling at a speed $v = 3/5$ undergoes a completely inelastic collision with an identical particle at rest (i.e., the two particles coalesce into one final particle during the collision). What is the speed of the single resulting body? What is its mass? Solve this problem both graphically and algebraically.

10.3 NEWTONIAN MOMENTUM IS NOT CONSERVED. Show that ordinary, newtonian momentum is not conserved in the example discussed in Sec. 10.3.

10.4 WE'RE UNDER ATTACK! A spaceship with rest mass m_0 is traveling with an x velocity $v_{0x} = +4/5$ in the frame of the earth. It collides with a photon torpedo (an intense burst of light) moving in the $-x$ direction relative to the earth. Assume that the ship's shields totally absorb the photon torpedo.

a The oncoming torpedo is measured by terrified observers on the ship to have an energy of $0.75m_0$. What is the energy of the photon torpedo in the frame of the earth? (*Hint:* How do the components of the four-momentum transform when we go from one inertial frame to another?)

b Use conservation of four-momentum to determine the velocity and mass (in terms of m_0) of the damaged ship after it absorbs the torpedo (as measured in the earth frame). Solve this problem both graphically and algebraically.

10.5 EMERGENCY PROCEDURES. A spaceship of mass m_0 is traveling through an uncharted region of deep space. Suddenly its sensors detect a black hole dead ahead. In a desperate attempt to stop the spaceship, the pilot fires the forward photon engines. These engines convert the mass-energy of matter-antimatter fuel entirely into photons, which are emitted in a tight beam in the direction of the ship's motion. Assuming that the spaceship has an initial speed $v = 3/5$ with respect to the black hole, what fraction of its mass m_0 must be converted into energy to bring the spaceship to rest with respect to the black hole? Solve this problem both graphically and algebraically. (*Hint:* Treat the emitted light as one big flash.)

10.6 TRAVELING TO THE STARS. As discussed in Sec. 10.7, the most efficient possible rocket engine would take matter and antimatter fuel, combine them in a controlled way, and focus the resulting light into a tight beam traveling away from the stern of the spaceship. Imagine that you want to design a spaceship using such an engine that can boost a payload of 25 metric tons (that is, 25×1000 kg) to a final cruising speed of 0.95. What must be the total mass of the ship at takeoff? [*Hint:* The ship can essentially be considered to be a particle of mass M at rest that decays into a big flash of light and a smaller particle (the payload) of known mass m traveling at a known speed v. Use conservation of four-momentum to determine M.]

10.7 TRAVELING TO THE STARS II. Consider further the rocket design problem in the previous exercise.

a Assume that you can find astronauts who are willing to travel for up to 50 years (as measured by their watches) on a round-trip to the stars. About how many light-years could the ship go out and return within that time constraint if it cruises at a speed of 0.95 and spends a negligible amount of time accelerating and decelerating? Is this very far compared to the galaxy as a whole?

b The takeoff mass calculated in Prob. 10.6 only included sufficient fuel to boost the payload to the cruising speed. For a complete round-trip, one would have to carry enough fuel to boost the payload to the cruising speed, decelerate it to rest at the destination, boost it to cruising speed again for the return trip home, and decelerate it upon reaching earth. How much fuel is required for a complete round-trip? Express your answer as a multiple of the payload mass m. (*Hint:* The answer is *not* four times the fuel required to boost the payload alone to the cruising speed, since the fuel required for all future boosts and decelerations must be boosted as well?)

c Comment on the practicality of visiting the distant stars using a rocket which must carry its own fuel.

10.8 TRAVELING TO THE STARS III. Derive a mathematical expression for the fuel-to-payload ratio for a perfect matter-antimatter rocket if it is to reach a final speed of v. Check that your formula reproduces the results of the example in Sec. 10.7.

10.9 LIGHT PRESSURE. Consider a freely floating mirror placed initially at rest in a laser beam. Imagine the mirror to be oriented directly facing the beam so that it reflects the beam back the way it came. Each flash of light that rebounds from the mirror has undergone a change in its momentum as a result of the change in the direction of its velocity. This means that the mirror will have to recoil a bit from each rebound to conserve four-momentum. Use conservation of four-momentum to estimate how much power the laser beam would have to have to accelerate a perfect 1-g mirror at a rate of 1 cm/s^2. Express your answer in watts (1 W = 1 J/s).

10.10 LIGHT PRESSURE. A flashlight emits a continuous beam of light forward. If the beam has a power of 5 W, with about how much force will the flashlight recoil backward against the hand of the person holding it? Is this going to be noticeable?

10.11 LIGHT SAILING. A method for getting around the difficulties discussed in Probs. 10.5 and 10.6 is to use light pressure to accelerate a payload. Imagine that you attach a perfect mirror to the back of your payload and then accelerate the payload by bouncing a powerful laser beam off the mirror. (This has the tremendous advantage that you do not have to carry the mass of the fuel or the rocket engine!) The lasers producing the beam could be massive things powered by solar energy, so neither the size, mass, nor power of these driving lasers is a significant limitation (at least in principle). Imagine that you wish to accelerate a 2000-kg scientific payload outward from the earth's orbit at a rate of 1 m/s^2 (at this rate, it would take about a year to reach 10 percent of the speed of light). Assume that the payload has been delivered to a point far enough from the earth so the earth's gravity is negligible (but *do not* ignore the gravity of the sun). How many watts of light would the driving laser have to produce?

11

APPLICATIONS TO PARTICLE PHYSICS

Everything that they have learned up to the age of eighteen is believed to be experience. Whatever they hear about later is theory and speculation.

Einstein, complaining about scientists with insufficient imagination[1]

11.1 INTRODUCTION

In this book we have been exploring the strange consequences of the principle of relativity and how these consequences imply that the universe is in fact very different from the newtonian picture of it that we build on the basis of our daily experience. The theory of special relativity is strange to us precisely because newtonian physics (which is based on *galilean* relativity) is quite adequate for describing and explaining *most* of the physics we encounter on a daily basis. While in this book we have imagined planes, trains, spaceships, runners, and the like that can travel at relativistic speeds, in fact it is impossible at present to boost any macroscopic object to more than a pathetically tiny fraction of the speed of light.

However, high-energy particle physics is one area of physics where special relativity is not only useful but absolutely essential. Because subatomic particles have such small masses, it is possible with current technology to accelerate them to speeds almost indistinguishably less than that of light, and this is done daily at particle accelerator facilities around the world. Even natural processes such as the decay of radioactive nuclei can produce particles moving at relativistic speeds.

[1]Quoted by E. H. Hutton in A. P. French (ed.), *Einstein: A Centenary Volume,* Cambridge, Mass.: Harvard, 1979, p. 177.

In the realm of high-energy particle physics, standard newtonian ideas are inadequate, but the ideas of special relativity, and most particularly the law of conservation of four-momentum, have been shown to be both descriptive and accurate (indeed, the frequency and precision with which the implications of special relativity are tested at the world's particle accelerators makes it one of the best-tested theories in all of physics). This is the arena, therefore, where special relativity is applied on a daily basis to practical, real-world problems.

The purpose of this chapter is to provide a short introduction to subatomic particle physics and to present some examples of how the law of conservation of four-momentum is applied in this context. Part of the point is to make it clear that special relativity is not just a beautiful theoretical speculation but has important *practical* and testable implications and represents one of the cornerstones on which the edifice of modern physics is constructed.

11.2 UNITS FOR PARTICLE PHYSICS

Let us first review a few things about the units we use to describe the four-momentum of an object. In the SR unit system, mass, momentum, and energy all have units of kilograms. Thus the fundamental unit of energy in the SR system is the *kilogram of energy.* To convert from this unit to joules ($1\ \text{J} \equiv 1\ \text{kg·m}^2/\text{s}^2$), the conventional SI unit of energy, one must multiply by *two* powers of the conversion factor c:

$$\begin{aligned} 1\ \text{kg}\ (\textit{of energy}) &= 1\ \text{kg}\left(\frac{2.998 \times 10^8\ \text{m}}{1\ \text{s}}\right)^2 = 8.988 \times 10^{16}\ \text{kg·m}^2/\text{s}^2 \\ &= 8.988 \times 10^{16}\ \text{J} \end{aligned} \tag{11.1}$$

Thus the kilogram of energy is a *lot* of energy, roughly equivalent to the energy output of a very large (1000-MW) electrical power plant over a time period of 2.8 years. This implies that a tiny amount of mass (i.e., rest energy) corresponds to a huge amount of energy in other forms. We will see that subatomic and nuclear reactions can convert a significant fraction of the reactants' rest energy into other forms of energy, thus producing prodigious amounts of such energy from a small amount of converted mass.

The fundamental unit of momentum in the SR system is the *kilogram of momentum.* To convert to the conventional SI units for momentum (kg·m/s), one must multiply by *one* power of the conversion factor c:

$$1\ \text{kg}\ (\textit{of momentum}) = 1\ \text{kg}\left(\frac{2.998 \times 10^8\ \text{m}}{1\ \text{s}}\right) = 2.998 \times 10^8\ \text{kg·m/s} \tag{11.2}$$

Again, the kilogram of momentum is a substantial amount of momentum, about equal to the momentum of a 30-ton truck traveling at roughly 22,000 mi/h.

The fact that mass, energy, and momentum are all measured in kilograms in the SR system reflects the fundamental unity of these quantities: the mass, energy, and momentum of an object are simply different aspects (specifically, the magnitude, time component, and the magnitude of the spatial components, respectively) of its four-momentum vector.

But we also see that the kilogram is an inconveniently large unit of mass, energy, and momentum when we are studying the dynamics of subatomic particles. For example, a proton moving at four-fifths the speed of light in the $+x$ direction has a mass, energy, and x momentum of about 1.67×10^{-27} kg, 2.78×10^{-27} kg, and 2.23×10^{-27} kg, respectively; it is inconvenient, to say the least, to work with numbers of this size. Particle physicists find it convenient to modify the SR system of units one step further and measure mass, momentum, and energy in terms of a different unit called the *electronvolt,* or eV, where 1 eV is defined to be the energy gained by a single electron as it passes through a potential difference of 1 volt. The conversion factor to joules is

$$1 \text{ eV} \equiv 1.602 \times 10^{-19} \text{ J} \tag{11.3a}$$

This is a very small unit of energy, well suited to the study of subatomic particle interactions. Through Eq. (11.1) we can link this unit to the SR unit of kilograms:

$$1 \text{ eV} = 1.602 \times 10^{-19} \text{ J}\left(\frac{1 \text{ kg}}{8.998 \times 10^{16} \cancel{\text{J}}}\right) = 1.782 \times 10^{-36} \text{ kg} \tag{11.3b}$$

Using this conversion factor, all masses, energies, and momenta can be expressed in electronvolts. For example, the mass of the electron is

$$m_e = 9.11 \times 10^{-31} \cancel{\text{kg}}\left(\frac{1 \text{ eV}}{1.782 \times 10^{-36} \cancel{\text{kg}}}\right) = 5.11 \times 10^5 \text{ eV} = 0.511 \text{ MeV} \tag{11.4}$$

11.3 SUBATOMIC PARTICLE PROPERTIES

Table 11.1 lists some subatomic particles and their properties, including masses given in units of MeV $\equiv 10^6$ eV (and provides sufficient information to do all the problems in this chapter). The superscript attached to the particle symbol in each case indicates the sign of the charge of the particle. Every particle has a corresponding antiparticle having opposite charge and the same mass and lifetime. The conventional notation for an antiparticle involves putting a bar over the symbol for the particle (for example, $\overline{\Sigma}^+$ is the *negatively* charged antiparticle corresponding to the Σ^+ particle), but in certain cases where there is no ambiguity, the bar is omitted and the charge superscript is reversed (for example, e^+, μ^+, π^-, K^-, p^-).

11.4 EXAMPLE: A MOVING ELECTRON

Problem In the laboratory frame, an electron is observed to move with such a speed that its relativistic kinetic energy is $K = 0.20$ MeV. How fast is the electron moving in that frame? What is the relativistic momentum p in that frame? If the electron is moving along the $+x$ axis, what is its four-momentum vector in that frame?

TABLE 11.1
SOME SUBATOMIC PARTICLE PROPERTIES*

Category	Name	Symbol	Antiparticle	Mass, MeV	Half-life, s
Field particles are carriers	Gluon	g	Same	0	Stable
of fundamental forces.	Photon	γ	Same	0	Stable
	W boson	W^+	W^-	81,000	$\approx 10^{-24}$
	Z boson	Z^0	Same	92,000	$\approx 10^{-24}$
Leptons are fairly lightweight	Neutrino†	ν	$\overline{\nu}$	0 (?)	Stable (?)
particles that do not part-	Electron	e^-	e^+	0.511	Stable
icipate in what is called the	Muon	μ^-	μ^+	105.66	1.524×10^{-6}
"strong" nuclear interaction.	Tau	τ^-	τ^+	1784	3.0×10^{-13}
Mesons are all construct-	Pi meson	π^0	Same	134.97	0.58×10^{-16}
ed of a quark and an anti-	(Pion)	π^+	π^-	139.57	1.804×10^{-8}
quark. There are actually	K meson	K^+	K^-	493.6	0.857×10^{-8}
two different kinds of K^0	(Kaon)	K^0	$\overline{K}^0$	497.7	$359 / 0.618 \times 10^{-10}$
particle, one with a longer	D meson	D^+	D^-	1869	7.4×10^{-13}
half-life than the other.	Psi	ψ	Same	3097	6.9×10^{-21}
	B meson	B^+	B^-	5278	9.1×10^{-13}
	Upsilon	Υ	Same	9460	9.0×10^{-21}
Baryons are all construct-	Proton	p^+	p^-	938.27	Stable? (> 10^{38})
ed of triplet of quarks.	Neutron	n	$\overline{n}$	939.57	621
The baryons listed here	Lambda	Λ^0	$\overline{\Lambda}^0$	1116	1.82×10^{-10}
are all constructed of	Sigma plus	Σ^+	$\overline{\Sigma}^+$	1189	0.554×10^{-10}
the combinations of the *u*,	Sigma zero	Σ^0	Σ^0	1193	5×10^{-20}
d, and *s* quarks (the	Sigma minus	Σ^-	$\overline{\Sigma}^-$	1197	1.03×10^{-10}
three lowest-mass quarks).	Xi zero	Ξ^0	$\overline{\Xi}^0$	1315	2.0×10^{-10}
	Xi minus	Ξ^-	$\overline{\Xi}^-$	1321	1.14×10^{-10}
	Omega minus	Ω^-	$\overline{\Omega}^-$	1672	0.57×10^{-10}

*Based on Particle Data Group, "Review of Particle Properties," *Phys. Lett.*, 204B, (April 1988), and other sources.
†There are actually three kinds of neutrinos, one corresponding to each kind of massive lepton.

Solution According to Eq. (9.24), the electron's total energy is given by

$$E = K + m = 0.20 \text{ MeV} + 0.51 \text{ MeV} = 0.71 \text{ MeV} \tag{11.5}$$

Knowing this and the mass of the electron allows us to calculate its relativistic momentum using Eq. (9.29):

$$m^2 = E^2 - p^2 \qquad \text{so} \qquad p^2 = E^2 - m^2$$

Thus

$$p = [(0.71 \text{ MeV})^2 - (0.51 \text{ MeV})^2]^{1/2} \approx 0.49 \text{ MeV} \tag{11.6}$$

Then, by Eq. (9.30) we have

$$v = \frac{p}{E} = \frac{0.49 \text{ MeV}}{0.71 \text{ MeV}} \approx 0.69 \tag{11.7}$$

Since the electron's motion is entirely along the $+x$ axis, its four-momentum is

$$\mathbf{P} = \begin{bmatrix} P_t \\ P_x \\ P_y \\ P_z \end{bmatrix} = \begin{bmatrix} E \\ +p \\ 0 \\ 0 \end{bmatrix} = \begin{bmatrix} 0.71 \text{ MeV} \\ 0.49 \text{ MeV} \\ 0 \\ 0 \end{bmatrix} \tag{11.8}$$

11.5 EXAMPLE: KAON DECAY

Problem A K^0 kaon at rest decays into two π^0 pions. Choose the $+x$ direction to be the direction of motion of *one* of the π^0 pions. In what direction does the *other* pion move? What are the velocities and kinetic energies of the two pions?

Solution In this case, the law of conservation of four-momentum implies that the sum of the product pions' four-momenta must equal the four-momentum of the original kaon. Since the kaon is at rest, Eq. (9.10*a*) implies that the time component of its four-momentum is equal to its mass ($M = 498$ MeV, according to Table 11.1), and Eqs. (9.10*b*) through (9.10*d*) imply that the spatial components of the kaon's four-momentum are equal to zero. Similarly, Eqs. (9.10*c*) and (9.10*d*) imply that $P_y = P_z = 0$ for the pion moving in the $+x$ direction, since it has no component of velocity in the y and z directions by hypothesis. Let us call the relativistic energy of this pion E_1 and its relativistic momentum p_1. The other pion we know nothing about yet: let us represent its relativistic energy by E_2 and the spatial components of its four-momentum by P_{2x}, P_{2y}, and P_{2z}. The mass of each pion produced by the decay is $m = 135$ MeV, according to Table 11.1. With these symbols defined, the law of conservation of four-momentum then requires that

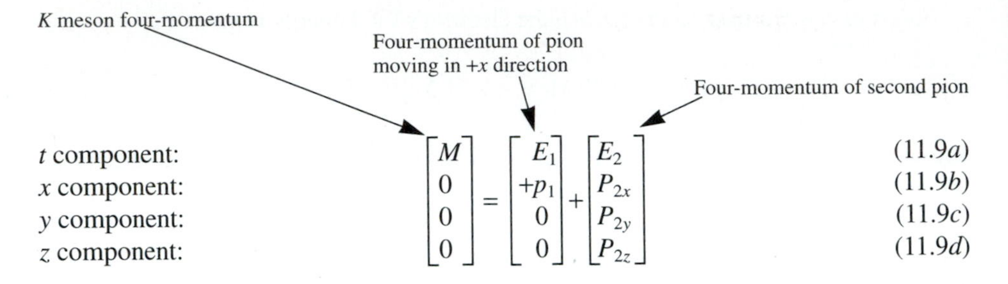

$$\begin{bmatrix} M \\ 0 \\ 0 \\ 0 \end{bmatrix} = \begin{bmatrix} E_1 \\ +p_1 \\ 0 \\ 0 \end{bmatrix} + \begin{bmatrix} E_2 \\ P_{2x} \\ P_{2y} \\ P_{2z} \end{bmatrix}$$

Equation (11.9*c*) reads $0 = 0 + P_{2y}$, implying that $P_{2y} = 0$. Similarly Eq. (11.9*d*) implies that $P_{2z} = 0$. Equation (11.9*b*) reads $0 = p_1 + P_{2x}$, implying that $P_{2x} = -p_1$. In short, the final three equations imply that the second pion moves in the $-x$ direction but that the relativistic momenta of the two pions are the same: $p_1 = |P_{1x}| = |P_{2x}| = p_2$. Now, both pions have the same mass m as well, so according to Eq. (9.29) we have

$$m^2 = E_1^2 - p_1^2 = E_2^2 - p_2^2 = E_2^2 - p_1^2 \tag{11.10a}$$

since $p_1 = p_2$. This implies that

$$E_1^2 = E_2^2 \Rightarrow E_1 = E_2 \tag{11.10b}$$

since energies are always positive. Therefore, the time component of conservation of four-momentum [Eq. 11.9*a*)] reduces in this case to

$$M = E_1 + E_2 = E_1 + E_1 \Rightarrow E_1 = \tfrac{1}{2}M = \tfrac{1}{2}(498\text{ MeV}) = 249\text{ MeV} \tag{11.10c}$$

The two pions thus have the same kinetic energy,

$$K = E_1 - m = 249\text{ MeV} - 135\text{ MeV} = 114\text{ MeV} \tag{11.11}$$

the same relativistic momentum,

$$m^2 = E_1^2 - p_1^2 \tag{11.12a}$$

$$m^2 \Rightarrow p_1 = p_2 = \sqrt{E_1^2 - m^2} = \sqrt{(249\text{ MeV})^2 - (135\text{ MeV})^2} \approx 209\text{ MeV} \tag{11.12b}$$

and the same speed [according to Eq. (9.30)],

$$v_1 = \frac{p_1}{E_1} = \frac{209\text{ MeV}}{249\text{ MeV}} = 0.839 \qquad v_2 = \frac{p_2}{E_2} = \frac{p_1}{E_1} = v_1 \tag{11.13}$$

though the two pions move in opposite directions along the x axis.

11.6 THE FOUR-MOMENTUM OF A PHOTON

I have mentioned before in this text that light can be understood as an electromagnetic wave: the wave nature of light can be easily demonstrated with diffraction experiments. But since the early part of this century, it has also been understood that light behaves under certain circumstances as if it were made up of tiny particles, called **photons**.

The experimental evidence for this particlelike aspect of light is beyond the scope of this text: such evidence is typically discussed in courses on modern physics. Suffice it to say that in subatomic particle interactions, the particle nature of light becomes its crucial aspect.

The link between these different ways of looking at light is the assertion (first made by Einstein in 1905) that each photon has an energy $E = h\nu$, where ν = the frequency of the light (thought of as a wave) and h is Planck's constant. The frequency ν has units of seconds^{-1} and in fact is equal to $1/\lambda$, where λ is the wavelength of the light in question (if λ is measured in the SR unit of seconds). Therefore, the energy of a photon of light is given by

$$E = h\nu = \frac{h}{\lambda} \qquad \text{(in SR units)} \tag{11.14a}$$

where

$$h = 7.37 \times 10^{-51}\ \text{kg·s} \tag{11.14b}$$

If E is expressed in energy units of electronvolts instead of the standard SR unit of kilograms, then

$$h = 4.14 \times 10^{-15}\ \text{eV·s} = 4.14\ \text{eV·fs} \tag{11.14c}$$

where 1 fs = 1 femtosecond $\equiv 10^{-15}$ s. For future reference, note that 1 fs of *distance* is

$$1\ \text{fs (distance)} = 2.998 \times 10^{-7}\ \text{m} = 299.8\ \text{nm} \tag{11.14d}$$

As discussed in Sec. 10.4, the only way we can create a self-consistent four-momentum vector for *anything* that carries energy at the speed of light is to assume that the object has zero mass. This means that photons must have *zero mass,* which in turn means that the photon's energy E and its relativistic momentum p must be equal:

$$0 = m^2 = E^2 - p^2 \Rightarrow E = p \qquad \text{(for a photon)} \tag{11.15}$$

The photon may not be the only particle that has zero mass. For many years a family of particles called **neutrinos** (which are produced by neutron decay, thermonuclear fusion, and many other particle decay and interaction processes) has been assumed to have zero mass. Recently, some doubts have been raised about this assumption, but even if neutrinos have a nonzero mass, experiments have shown that their mass must be at least 100,000 times smaller than that of an electron (the lightest known particle with definitely nonzero mass), which is essentially zero on the scale of masses and energies involved in particle reactions that we will consider. There are other theoretically predicted particles (**gravitons, gluons,** etc.) that may have zero mass but are not directly observable at present; we will not consider examples involving these particles.

We will, however, work with both photons and neutrinos in the examples and problems below. The conventional symbol for the photon is the Greek letter γ *(gamma);* the symbol for the neutrino is the Greek letter ν *(nu).* Both particles have zero charge. For the sake of this text, you should assume that both these kinds of particles have exactly zero mass. The crucial thing to know about such particles is that *the energy of a massless particle is equal to it relativistic momentum,* as implied by Eq. (11.15). Such particles will *always* travel at exactly the speed of light: if $E = p$, then we have

$$v = \frac{p}{E} = 1 \tag{11.16}$$

We will find these bits of knowledge very useful in the examples and problems that follow.

11.7 EXAMPLE: COMPTON SCATTERING

Problem Certain radioactive substances radiate high-energy photons called *gamma rays.* The energy of a *single* such photon can be measured with modern equipment. Imagine that such a photon with energy E_0 collides with an electron of mass m at rest and rebounds from it back along the direction from which it came. What is the energy of the rebounding photon? See Fig. 11.1.

Solution Figure 11.2 shows an energy-momentum diagram of such a collision, taking the energy of the photon to be equal to the electron's mass m for the sake of concreteness. This diagram clearly shows that $E < E_0$, that is, the photon's energy after the collision is smaller than its original energy (it actually looks that if we choose $E_0 = m$, E turns out to be a bit larger than $m/3$).

The solution can be found algebraically as well. Since the electron is at rest before the collision, its original x momentum is zero and its original energy is equal to its

FIGURE 11.1
A photon with energy E_0 collides with an electron at rest. What is the energy E of the rebounding photon?

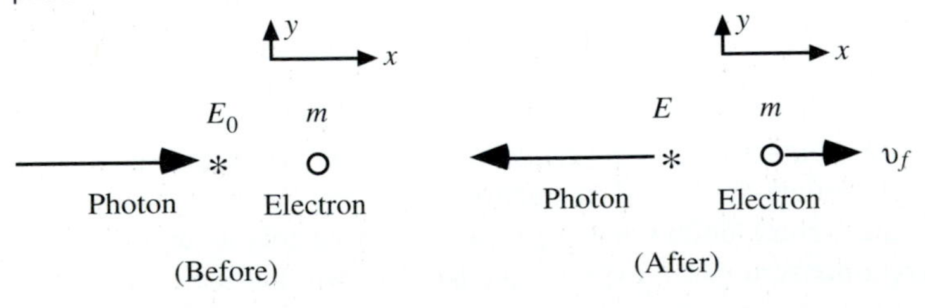

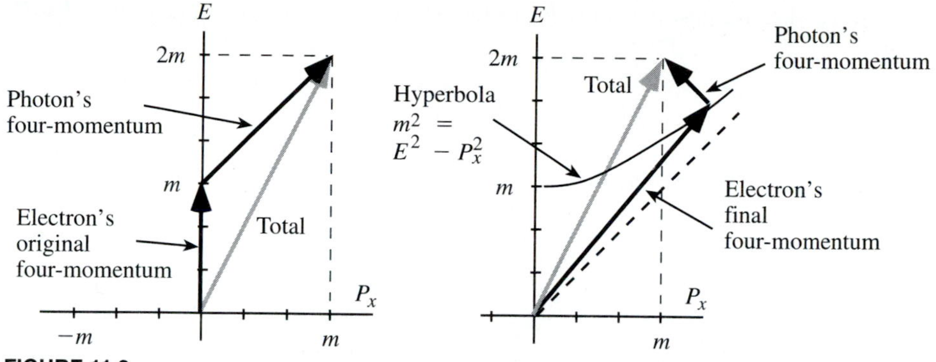

FIGURE 11.2
(*a*) Energy-momentum diagram for the four-momenta of the photon and electron before the collision. Note that the slope of the photon four-momentum must = 1. (*b*) The electron's final four-momentum will have its tip somewhere on the hyperbola shown. The final photon four-momentum must have a slope of −1 if it is moving in the −*x* direction. So draw a line of slope −1 from the tip of the total four-momentum vector until it intersects the hyperbola: this is where the tip of the electron's final four-momentum arrow must be located.

mass m. Since $p = E$ for photons and E is always positive, the original photon's x momentum must be $+E_0$ and the recoiling photon's x momentum must be $-E$.

Let E_f and P_{fx} be the energy and x momentum of the electron *after* the collision. The law of conservation of four-momentum implies that

$$\begin{array}{l} t \text{ component:} \\ x \text{ component:} \\ y \text{ component:} \\ z \text{ component:} \end{array} \qquad \begin{bmatrix} E_0 \\ +E_0 \\ 0 \\ 0 \end{bmatrix} + \begin{bmatrix} m \\ 0 \\ 0 \\ 0 \end{bmatrix} = \begin{bmatrix} E \\ -E \\ 0 \\ 0 \end{bmatrix} + \begin{bmatrix} E_f \\ P_{fx} \\ P_{fy} \\ P_{fz} \end{bmatrix} \tag{11.17a}$$

The bottom two equations tell us that $P_{fy} = P_{fz} = 0$, meaning that the recoiling electron moves along the x axis with relativistic momentum $p_f = |P_{fx}|$.

In this problem, E_0 and m are known (we will assume that the original photon energy is given, and we can look up the electron's mass). So these equations provide two equations in the *three* unknowns E, E_f, and p_f. But we also know that

$$m^2 = E_f^2 - p_f^2 \tag{11.17b}$$

since the magnitude of the recoiling electron's four-momentum must be equal to the electron's mass. This provides the third equation that we need to solve for the three unknowns.

According to the x-component part of Eq. (11.17*a*),

$$P_{fx} = E_0 + E = +p_f \tag{11.18a}$$

(since E_0 and E are positive, so is P_{fx}.) Plugging this into (11.17*b*) and solving for E_f, we get

$$E_f^2 = m^2 + p_f^2 = m^2 + (E_0 + E)^2 \tag{11.18b}$$

Plugging this into the *t*-component part of Eq. (11.17*a*) to eliminate E_f, we get

$$E_0 + m = E + \sqrt{(E^0 + E)^2 + m^2} \tag{11.18c}$$

or

$$(E_0 + m - E)^2 = (E_0 + E)^2 + m^2 \tag{11.18d}$$

Our task now is basically to solve for E. Writing out the squares in the last equation, we get

$$E_0^2 + 2mE_0 - 2E_0E - 2mE + m^2 + E^2 = E_0^2 + 2E_0E + E^2 + m^2 \tag{11.18e}$$

Canceling E_0^2, E^2, and m^2 from both sides, we get

$$+2mE_0 - 2E_0E - 2mE = 2E_0E \qquad \text{or} \qquad 2mE_0 = 4E_0E + 2mE \tag{11.18f}$$

or

$$(2E_0 + m)E = mE_0 \qquad \text{implying that} \qquad E = \frac{mE_0}{2E_0 + m} \tag{11.18g}$$

When $E_0 = m$ (the case shown in Fig. 11.2), $E = m/3$, roughly equal to the result that we found graphically on the diagram.

An experiment to test this formula can be performed as follows. Photons with a known energy are directed against a metal block. Some of these photons collide with free electrons in the metal and bounce back. The energy of these rebounding photons can be measured. If a range of initial photon energies is used, the energies of the rebounding photons can be determined as a function of the original photon energy E_0.

This experiment was first performed by Arthur Compton in 1920 and has been repeated many times since. The results are in complete agreement with Eq. (11.18*g*). This is not only a vindication of the idea of conservation of four-momentum and the assumptions that we have made about how to handle photon four-momentum, but it represents compelling evidence for the idea that light can really be treated as if it were made up of *photons* that rebound from electrons as if they were billiard balls! Historically, Compton's work was what finally convinced the physics community that the photon model of light was something more than a theorist's speculation.

11.8 EXAMPLE: PAIR PRODUCTION

At the level of subatomic particle physics, an **inelastic** collision is one that converts energy of motion into energy of mass by the creation of new particles.[1] Fundamental particle theory predicts that when new particles are created out of kinetic energy, they must be created in pairs: every particle must be accompanied by its antiparticle. For example, if an electron is created by an inelastic collision, an antielectron (a *positron*) must also be created.

Now a single gamma-ray photon can carry more energy than the mass energy of an electron-positron pair. Nonetheless, the reaction

$$\gamma \rightarrow e^+ + e^- \tag{11.19}$$

is not observed to occur. In fact, such a process *cannot* conserve four-momentum, as I will now demonstrate. Let us take the direction of motion of the photon to define the $+x$ direction. Let the energies of the gamma-ray photon, the positron, and the electron be E_0, E_1, and E_2, respectively: conservation of the time component of the four-momentum thus requires

$$E_0 = E_1 + E_2 \tag{11.20}$$

Equation (9.29) implies that for *any* particle with nonzero mass (such as the electron or the positron in the case under consideration),

$$m^2 = E^2 - P_x^2 - P_y^2 - P_z^2 \Rightarrow P_x^2 = E^2 - P_y^2 - P_z^2 - m^2 < E^2 \Rightarrow |P_x| < E \tag{11.21}$$

since E is always positive. Plugging this into Eq. (11.20), we see that

$$E_0 > P_{1x} + P_{2x} \tag{11.22}$$

Now, the photon's relativistic momentum is equal to E_0, because it is massless. Since the photon is moving in the $+x$ direction by hypothesis, its x component of momentum is $P_x = +E_0$. Therefore, conservation of the x component of four-momentum requires that

$$E_0 = P_{1x} + P_{2x} \tag{11.23}$$

Do you see that this leads to a contradiction? Equation (11.23) implies that E_0 is strictly *equal* to $P_{1x} + P_{2x}$. Equation (11.22) implies that E_0 is strictly *greater than*

[1]Analogously, the Compton scattering process just described would be considered an **elastic** collision: no mass-energy was created or destroyed, implying that the kinetic energy of the system was conserved. The particle decay process described in Sec. 11.5 would be considered a **superelastic** process, in that the final configuration of the system has more kinetic energy than the original configuration: some mass-energy was converted into kinetic energy.

$P_{1x} + P_{2x}$. These equations cannot be simultaneously true, and yet they *must* be simultaneously true for four-momentum to be conserved. The conclusion is that four-momentum *cannot* be conserved for the process described in (11.19), and thus the process cannot occur.

This assertion is even more vividly illustrated by the energy-momentum diagram shown in Fig. 11.3. This diagram shows that one of the pair of particles created from the photon would have to be moving faster than the speed of light if the total four-momentum of the created pair is to be equal to the original four-momentum of the photon.

It *is* possible, however, to create an electron-positron pair from a single gamma-ray photon by means of the reaction

$$\gamma + e^{-}\ (\text{at rest}) \rightarrow e^{-} + e^{-} + e^{+} \tag{11.24}$$

In this interaction, the gamma-ray photon collides with an electron at rest and disappears, leaving an electron-positron pair in its place. The presence of the extra electron in this interaction enables the four-momentum of the system to be conserved, as the worked example below demonstrates.

Problem Consider the collision process described by (11.24). Define the $+x$ direction to be the direction of motion of the original gamma-ray photon. For simplicity's sake, imagine that the product particles in this reaction move together with the same velocity in the $+x$ direction after the collision. Show that four-momentum can be conserved in such a process if the gamma-ray photon has a certain energy and calculate this energy.

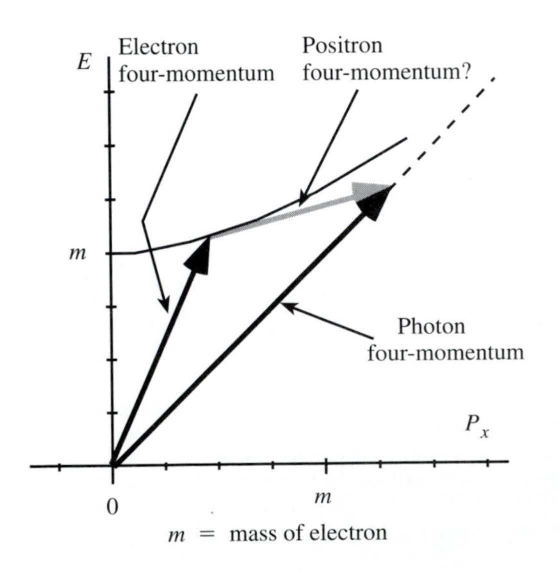

FIGURE 11.3
This diagram illustrates that it is impossible for a single photon to produce an electron-positron pair. The four-momentum vector of the initial photon has a slope of 1 on this diagram, since the photon moves at the speed of light. The created electron must move at a speed <1, since it has nonzero mass, so the slope of its four-momentum will be less than 1. Conservation of four-momentum requires that the four-momenta of the electron and positron add up to that of the photon. But the only way for this to happen is if the positron's four-momentum has a slope greater than 1, implying that the positron is moving faster than light. This is impossible, so the decay process is impossible.

Solution Let the unknown energy of the gamma-ray photon be E_0. Since it is massless and moving in the $+x$ direction, its four-momentum will be $[P_t, P_x, P_y, P_z] = [E_0, +E_0, 0, 0]$. The original electron at rest has a four-momentum of $[m, 0, 0, 0]$, where m is the mass of the electron. The three product particles all have the same velocity by hypothesis, and they also have the same mass, so their total four-momentum will be the same as the four-momentum of a single particle of mass $3m$ traveling at their joint velocity. Let us denote the total four-momentum vector of these three particles by $[E_3, P_{3x}, P_{3y}, P_{3z}]$ and note that according to Eq. (9.29),

$$(3m)^2 = E_3^2 - P_{3x}^2 - P_{3y}^2 - P_{3z}^2 \tag{11.25}$$

The law of conservation of four-momentum then requires that

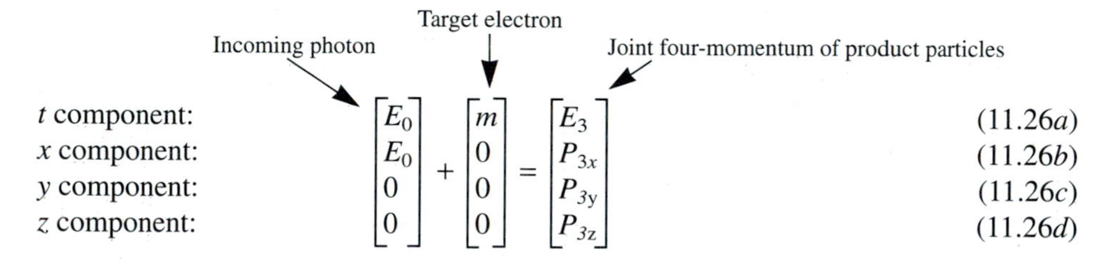

$$\begin{bmatrix} E_0 \\ E_0 \\ 0 \\ 0 \end{bmatrix} + \begin{bmatrix} m \\ 0 \\ 0 \\ 0 \end{bmatrix} = \begin{bmatrix} E_3 \\ P_{3x} \\ P_{3y} \\ P_{3z} \end{bmatrix} \qquad \begin{matrix} (11.26a) \\ (11.26b) \\ (11.26c) \\ (11.26d) \end{matrix}$$

Equations (11.26*c*) and (11.26*d*) imply that if the product particles move together after the collision, they must move in the $+x$ direction. Equation (11.26*b*) implies that $P_{3x} = E_0$. Plugging this into Eq. (11.25), we find that

$$(3m)^2 = E_3^2 - E_0^2 \quad \Rightarrow \quad E_3 = \sqrt{E_0^2 + (3m)^2} \tag{11.27}$$

Plugging this into Eq. (11.26*a*), we get

$$\begin{aligned} E_0 + m = \sqrt{E_0^2 + (3m)^2} \quad &\Rightarrow \quad E_0^2 + 2E_0 m + m^2 = E_0^2 + 9m^2 \\ &\Rightarrow \quad 2E_0 m = 8m^2 \\ &\Rightarrow \quad E_0 = 4m \end{aligned} \tag{11.28}$$

We see that four-momentum can be conserved in this process if the original photon has an energy of $4m \approx 4(0.511 \text{ MeV}) = 2.04$ MeV. This is twice as much energy as one might think would be required to produce the rest mass of the new particles, but a certain amount of the energy of the photon *must* be channeled into kinetic energy of the product particles if the four-momentum of the entire system is to be conserved.

11.9 EXAMPLE: THE DOPPLER SHIFT REVISITED

In Sec. 5.10, the Doppler shift formula was derived by considering a light wave to be a series of electromagnetic "pulses" separated by a definite period of time. In this section, I will rederive the Doppler shift formula, treating light as a conglomeration of photons and applying the four-vector transformation law to the photons' four-momenta.

Problem Imagine a light source that moves at a constant velocity β in the $+x$ direction with respect to an inertial frame (which we will take to be the Home Frame) and emits a beam of monochromatic (single-wavelength) light in the direction opposite to its motion. An observer at rest with respect to the source measures the light to have a certain wavelength λ_0. What is the wavelength λ of the light as observed in the Home Frame?

Solution Call the frame of the source the Other Frame, and imagine it to be in standard orientation with respect to the Home Frame. According to the photon picture of light, the light emitted by the source can be considered to consist of a large number of photons, all of which have the same energy $E' = h/\lambda_0$ [according to Eq. (11.14*a*)] in the Other Frame. Because each photon is massless and travels in the $-x$ direction, the four-momentum of any given photon in the Other Frame will have the following form: $[P'_t, P'_x, P'_y, P'_z] = [E', -E', 0, 0]$. Using the inverse Lorentz transformation equation for the time component of the four-momentum [see Eq. (9.12*a*)], we can determine the energy E of this photon as measured in the Home Frame:

$$E = P_t = \frac{P'_t + \beta P'_x}{\sqrt{1-\beta^2}} = \frac{E' - \beta E'}{\sqrt{1-\beta^2}} = E'\frac{1-\beta}{\sqrt{1-\beta^2}} = E'\sqrt{\frac{1+\beta}{1-\beta}} \tag{11.29}$$

Since Eq. (11.14*a*) applies to the Home Frame as well as the Other Frame, the relationship between the energy E and the wavelength of the light λ as measured in the Home Frame is simply $E = h/\lambda$. This means that

$$\frac{\lambda}{\lambda_0} = \frac{h/\lambda_0}{h/\lambda} = \frac{E'}{E} = \frac{E'}{E'\sqrt{\dfrac{1-\beta}{1+\beta}}} = \sqrt{\frac{1+\beta}{1-\beta}} \tag{11.30}$$

which is precisely the Doppler shift formula for wavelengths given by Eq. (5.25) (note that the x velocity of the object relative to the Home Frame is equal to β in this case). Thus Eq. (11.14*a*), the assumption that photons have zero mass, and the transformation law for the components of a four-vector conspire together to produce the same result for the Doppler shift that an entirely different derivation based on the wave model of light does!

11.10 PARTING COMMENTS

The list of examples one could discuss is endless: the principle of relativity has many more exciting and unusual consequences that might be explored. But books (if not lists) must end, and I would like to end this one with some suggestions as where the interested reader might go from here.

Any good college library will have many books on special relativity, several of them written at a more sophisticated and comprehensive level than this book. Recent books will generally present a more powerful and up-to-date picture of relativity than old classics. I particularly recommend E. F. Taylor and J. A. Wheeler's *Spacetime Physics,* now available in its second edition (New York: Freeman, 1992) for its modern perspective and interesting physics. The exercises in this book are especially challenging and mind expanding, ranging in difficulty and scope from the freshman to the graduate level.

One important consequence of the principle of relativity that has not been presented in this text is the fact that relativity lies behind the unity of electricity and magnetism: the phenomenon of magnetism turns out to be simply a relativistic manifestation of Coulomb's law! The argument for this is beautifully presented by Edward Purcell in chap. 5 of his text *Electricity and Magnetism* (The Berkeley Physics Course, vol. 2, New York: McGraw-Hill, 1965).

Almost every issue of the *American Journal of Physics* has an article concerning the foundations or applications of special relativity. Some recent articles, for example, include:

S. P. Boughn, "The Case of the Identically Accelerated Twins," *Am. J. Phys.*, **57** (9), September 1989.

K. G. Suffern, "The Apparent Shape of a Rapidly Moving Sphere," *Am. J. Phys.*, **56** (8), August 1988.

R. E. Gibbs, "Photographing a Relativistic Meterstick," *Am. J. Phys.*, **48**(12), December 1980.

J. M. McKinley and P. Doherty, "The Search for the Starbow: The Appearance of the Starfield from a Relativistic Spaceship," *Am. J. Phys.*, **47**(4), April 1979.

R. Perrin, "The Twin Paradox: A Complete Treatment from the Point of View of Each Twin," *Am. J. Phys.*, **47**(4), April 1979.

M. S. Greenwood, "Use of Doppler-Shifted Light Beams to Measure Time During Acceleration" [a study of the twin paradox], *Am. J. Phys.*, **44**(3), March 1976.

This journal is the place to go if one wants to keep abreast of the latest ideas in undergraduate-level relativity theory.

Ultimately, special relativity leads to general relativity, a theory that uses the metaphor of spacetime geometry to present a wholly new and exciting perspective on gravity. The full study of general relativity requires some mathematical sophistication: one must be able to handle arbitrary coordinate systems on curved surfaces in more dimensions than one can easily visualize. Nevertheless, *Flat and Curved Spacetimes* by G. F. R. Ellis and R. M. Williams (New York: Oxford, 1988) provides some insights in general relativity at roughly the level of this text, and B. F. Schutz, *A First Course in General Relativity* (New York: Cambridge, 1985), provides one of the best introductions I am aware of to general relativity at the advanced undergraduate level.

That so much can spring from the principle of relativity is wonderful (if problematic for the textbook writer). I heartily encourage all to continue the exploration!

PROBLEMS

11.1 FINDING THE FOUR-MOMENTUM. A π^0 pion has a relativistic energy E = 281 MeV. Show that its speed $\approx$ 0.877, and find the magnitude of its relativistic three-momentum p.

11.2 FINDING THE FOUR-MOMENTUM. A proton is measured to have a kinetic energy of 2.1 GeV. What is its speed? Find the magnitude of its relativistic three-momentum.

11.3 FINDING THE MASS. A certain particle has a total energy of 0.25 MeV and a relativistic three-momentum of magnitude 0.20 MeV. Find its mass in MeV.

11.4 FINDING THE MASS. A certain particle is measured to have a relativistic kinetic energy of 1.6 MeV and a three-momentum magnitude of 0.85 MeV. Find its mass in MeV.

11.5 RELATIVISTIC AND NONRELATIVISTIC KINETIC ENERGY. How fast does an object have to be moving before its relativistic kinetic energy K (as defined in Eq. 9.24) differs from its newtonian kinetic energy $\frac{1}{2}mv^2$ by 0.01 percent?

11.6 THE DECAY OF THE π^0 MESON. A π^0 meson ordinarily decays into two photons. Consider a π^0 meson traveling at a speed v = 4/5 in the $+x$ direction which decays into two photons, one moving in the $+x$ direction with energy E_1 and another moving in the $-x$ direction with energy E_2.

a Find the photon energies E_1 and E_2.

b Find the wavelengths of these photons and compare with the wavelength of visible light (400 to 700 nm).

11.7 THE DECAY OF THE Ξ^0. A Ξ^0 particle ordinarily decays into a Λ^0 particle and a π^0 particle. If the Ξ^0 is at rest in the lab frame when it decays, what are the velocities and kinetic energies of the product particles in that frame? (*Hint:* Define the $+x$ direction to be the direction of motion of one of the product particles.)

11.8 THE DECAY OF THE π^-. A π^- (pion) ordinarily decays into a μ^- (muon) and a neutrino. Assume that a pion is at rest in the lab frame when it decays, and define the $+x$ direction to be the direction of motion of the muon after the decay. What is the speed of the muon after the decay? What is the energy of the neutrino?

11.9 PAIR ANNIHILATION. An electron and a positron will annihilate each other (producing two photons) if they come close enough together: $e^- + e^+ \rightarrow \gamma + \gamma$. (This process must produce two photons instead of one if four-momentum is to be conserved.) Imagine that the electron-positron pair are essentially at rest in the Home Frame when the annihilation occurs. Define the $+x$ axis of this frame to be the direction of motion of one of the outgoing photons.

a Use the law of conservation of four-momentum to find the energies and relativistic momenta (both magnitude and direction) of the outgoing photons as observed in the Home Frame.

b Imagine that this process is observed in an Other Frame moving with β = 4/5 in the $+x$ direction with respect to the Home Frame. Use the Lorentz transformation equations to find the four-momenta of all the particles involved in this reaction as measured in this Other Frame.

c Use the results of part b to demonstrate that four-momentum is also conserved in the Other Frame, as required by the principle of relativity.

11.10 INTERACTION OF FICTITIOUS PARTICLES. A neutral bozon of mass m traveling with $v = 4/5$ in the $+x$ direction strikes a snoozon of mass $M = 3m$ at rest. The interaction of these particles produces a previously undiscovered particle (which we will call a "rayon") and a photon with energy $2m$. Assume that the photon is emitted in the $-x$ direction.

a What is the four-momentum of the rayon?

b Use the result of part *a* to find the mass, velocity, and kinetic energy of the rayon. Express the mass and kinetic energy as multiples of the mass m of the bozon.

11.11 ACCELERATOR PHYSICS. An accelerator produces a beam of π^- pions with a total energy of 620 MeV. This beam is focused on a target of liquid hydrogen. Some of the pions undergo collisions with the nuclei of the hydrogen atoms (which are simply protons).

a What is the velocity of the pions in the pion beam?

b Argue that the process $\pi^- + p^+ \rightarrow K^0 + \Lambda^0$ is physically impossible with a pion beam having this energy.

c Argue that the process $\pi^- + p^+ \rightarrow \pi^0 + n$ *is* physically possible. Imagine that we define the $+x$ direction to be the direction in which the π^- is moving before the interaction, and imagine that the neutron *(n)* moves in the $+x$ direction after the interaction. Describe in detail (setting up all the relevant equations) how you would find the energies and three-momenta of the π^0 and neutron. (You do not need to actually solve these equations; just describe the detailed process that you would go through to solve your equations for the requested values.)

11.12 THE DECAY OF THE NEUTRON. A free neutron will decay with a half-life of about 10.4 min into a proton, an electron, and an antineutrino. What is the maximum possible energy of the emitted electron in this case? What is the minimum energy?

CONVERSION OF EQUATIONS TO SI UNITS

If you can't beat 'em, join 'em.

Anonymous

A.1 WHY USE THE SR UNIT SYSTEM?

The equations of special relativity are greatly simplified when one uses SR units to measure distance, as we have seen. But the purpose of using SR units is not merely to simplify a few equations: using such units also vividly draws one's attention to the connections special relativity makes between quantities that were previously considered to be fundamentally distinct. For example, special relativity teaches us that energy, momentum, and mass are in fact different aspects of the same basic quantity, the four-momentum. It is not merely *convenient* to measure the time component E, the spatial components P_x, P_y, P_z, and the magnitude m of this four-vector in the same units, it is fundamentally *appropriate* as well. Similarly, the basic metaphor of spacetime geometry that lies at the root of both special and general relativity is obscured if one insists on using different units to measure time and distance.

Nonetheless, this choice of units does lead to complications when one tries to apply the ideas presented in this book to practical situations, since practicing physicists in their everyday work still use traditional units to describe quantities. It is important to be able to use the simple and beautiful equations in this book in situations where the quantities in question are expressed in traditional units. Fortunately, it is straightforward to convert equations appearing in this text to equivalent equations involving quantities measured in SI units. The purpose of this appendix is to describe an easy method for doing this.

A.2 CONVERSION OF BASIC QUANTITIES

The SR unit system as defined in Chap. 2 differs from the SI unit system only in the substitution of the *second* for the *meter* as the basic unit of distance. Mass and time thus have the same units in the SI and SR systems and need no conversion. The most important quantities that are affected by the shift in units as one changes systems are *distance, velocity, energy,* and *momentum.* We also should consider what happens to values of the universal constants c and h as we change unit systems.

To help keep things straight in what follows, let me denote quantities measured in SR units with an "(SR)" subscript; for example, the speed of an object in SR units will be written in this appendix as $v_{(SR)}$, the energy of an object in the SR unit of kilograms will be written $E_{(SR)}$, and so on. Quantities without this subscript will be assumed to be measured in SI units.

In Chap. 2, the general rule for converting SI quantities to SR quantities was to multiply the SI quantity by the appropriate factor of c that leads to the correct SR dimensions. Let us apply this rule to the quantities of interest listed above.

Distance in the SR system is measured in seconds. Distance in the SI system is measured in meters. To convert an SI distance x to an SR distance $x_{(SR)}$, we must divide x by one factor of c (in meters per second): $x_{(SR)} = x/c$. Energy has units of kilograms in the SR system; it has units of joules = kg · m^2/s^2 in the SI system. To convert from E (in joules) to $E_{(SR)}$ (in kilograms), we must divide E by two powers of c: $E_{(SR)} = E/c^2$. Planck's constant has units of energy·time. In the SR system, energy is measured in kilograms instead of joules, so again we have to divide the SI version of Planck's constant by two powers of c to get the correct SR units: that is, $h_{(SR)} = h/c^2$. Conversion equations involving velocity and momentum can be derived in a similar manner. The results are summarized in Table A.1.

TABLE A.1
SI UNIT EQUIVALENTS FOR SR UNIT QUANTITIES

Quantity	SR symbol	SI equivalent
Time	$t_{(SR)}$	t
Distance	$x_{(SR)}$	x/c
Speed (of object)	$v_{(SR)}$	v/c
Speed (of frame)	$\beta_{(SR)}$	β/c
Mass	$m_{(SR)}$	m
Momentum	$p_{(SR)}$	p/c
Energy	$E_{(SR)}$	E/c^2
Speed of light	1	c
Planck's constant	$h_{(SR)}$	h/c^2

A.3 CONVERTING SR UNIT EQUATIONS TO SI UNIT EQUATIONS

The trick for converting equations from the SR to the SI system is now very simple: You simply replace the SR quantities in an equation by the SI equivalents given in Table A.1. For example, consider the metric equation

$$\Delta s^2_{(SR)} = \Delta t^2_{(SR)} - \Delta x^2_{(SR)} - \Delta y^2_{(SR)} - \Delta z^2_{(SR)}$$

Both $\Delta s_{(SR)}$ and $\Delta t_{(SR)}$ have time units and so have the same value in both systems. But Δx, Δy, Δz are measured in distance units in the SI system, and therefore $\Delta x_{(SR)} = \Delta x/c$, $\Delta y_{(SR)} = \Delta y/c$, and $\Delta z_{(SR)} = \Delta z/c$. The metric equation in SI units is thus

$$\Delta s^2 = \Delta t^2 - \left(\frac{\Delta x}{c}\right)^2 - \left(\frac{\Delta y}{c}\right)^2 - \left(\frac{\Delta z}{c}\right)^2 \tag{A.1}$$

As another example, consider Eq. (11.14*a*) in the text, which gives the energy of a photon in terms of its wavelength: $E_{(SR)} = h_{(SR)}/\lambda_{(SR)}$. Both energy and Planck's constant gain a factor of c^{-2} when switching from SR to SI units: this factor divides out in the equation above. The wavelength λ, on the other hand, has units of meters in the SI system but seconds in the SR system, so $\lambda_{(SR)} = \lambda/c$. The equation thus becomes $E = hc/\lambda$ in SI units.

Table A.2 lists some of the important equations in the text and their SI equivalents. In many of the cases illustrated in Table A.2, the SI equations are simply found by substituting the SI unit equivalents from Table A.1 for the SR unit quantities in the equation from the text. However, in some cases, the SI unit equations have been further simplified by dividing out common factors of c. For example, the SR unit version of the equation giving the magnitude of the relativistic three-momentum of a particle in terms of its speed reads

$$p_{(SR)} = \frac{m_{(SR)}v_{(SR)}}{\sqrt{1-(v^2_{(SR)})}} \tag{A.2}$$

If we simply perform the substitutions called for in Table A.1, we get

$$\frac{p}{c} = \frac{m(v/c)}{\sqrt{1-(v/c)^2}} \tag{A.3}$$

The equation can be made prettier, however, by multiplying through by c:

$$p = \frac{mv}{\sqrt{1-(v/c)^2}} \tag{A.4}$$

This is the simplified equation given in Table A.2. Many of the equations in the right-hand column of the table have been simplified in this manner.

TABLE A.2
SOME IMPORTANT EQUATIONS FROM THE TEXT AND THEIR SI EQUIVALENTS

Equation	SR version	SI equivalent
Metric	$\Delta s^2 = \Delta t^2 - \Delta x^2 - \Delta y^2 - \Delta z^2$	$\Delta s^2 = \Delta t^2 - \dfrac{\Delta x^2 + \Delta y^2 + \Delta z^2}{c^2}$
Proper time	$d\tau = dt\sqrt{1 - v^2}$	$d\tau = dt\sqrt{1 - (v/c)^2}$
Lorentz transformations	$\gamma = \dfrac{1}{\sqrt{1-\beta^2}}$ $t' = \gamma(t - \beta x)$ $x' = \gamma(-\beta t + x)$	$\gamma = \dfrac{1}{\sqrt{1-(\beta/c)^2}}$ $t' = \gamma(t - \beta x/c^2)$ $x' = \gamma(-\beta t + x)$
Lorentz contraction	$L = L_R\sqrt{1-\beta^2}$	$L = L_R\sqrt{1-(\beta/c)^2}$
Velocity addition (x component)	$v_x = \dfrac{v'_x + \beta}{1 + \beta v_x}$	$v_x = \dfrac{v'_x + \beta}{1 + \beta v_x/c^2}$
Energy in terms of speed	$E = \dfrac{m}{\sqrt{1 - v^2}}$	$E = \dfrac{m}{\sqrt{1 - (v/c)^2}}$
Relativistic momentum in terms of speed	$p = \dfrac{mv}{\sqrt{1 - v^2}}$	$p = \dfrac{mv}{\sqrt{1 - (v/c)^2}}$
Four-magnitude of four-momentum	$m^2 = E^2 - p^2$	$(mc^2)^2 = E^2 - (pc)^2$
Velocity-momentum-energy relationship	$v = \dfrac{p}{E}$	$\dfrac{v}{c} = \dfrac{pc}{E}$
Photon energy	$E = h\nu = h/\lambda$	$E = h\nu = hc/\lambda$

Most of the applications of special relativity are in nuclear and particle physics. Physicists in these fields typically focus on *energy* as the most important dynamic quantity. Because this practice is so common, I have in the last chapter of this text implicitly modified the SR unit system described in Chap. 2 by measuring energies, momenta, and masses in *energy* units (typically electronvolts) instead of the announced standard *mass* unit of the kilogram. Note that in the SR equations dealing with four-momentum quantities (e.g., the last five equations in Table A.2), it does not matter what units one uses to express the quantities m, p, and E as long as one uses the *same* units for these quantities. The only important difference is that Planck's constant h must be expressed in units of energy·time instead of mass·time: see Eqs. (11.14b) and (11.14c) in the text.

PROBLEMS

A.1 Verify that $p_{(SR)} = p/c$, as claimed in Table A.1.

A.2 Convert the equation describing the Doppler shift [Eq (5.25) in the text] to its equivalent expressed in terms of SI units.

A.3 Convert the equation describing the relativistic kinetic energy of a particle [Eq. (9.24) in the text] to its equivalent in terms of SI units.

A.4 The equation for the spacetime separation between two events is

$$\Delta\sigma_{(SR)} = \Delta d^2_{(SR)} - \Delta t^2_{(SR)}$$

Since $\Delta\sigma$ is actually directly measured using a measuring stick as opposed to a clock, it would make sense to express its value in meters instead of seconds if one is going to use the SI unit system. With this in mind, what would be the equivalent of this equation expressed in terms of SI units?

A.5 What are the units of the universal gravitational constant $G_{(SR)}$ in the SR unit system? [*Hint:* The SI units of G can be inferred from Newton's law of universal gravitation: $F_g = Gm_1m_2/(r_{12})^2$.] Derive an equation expressing $G_{(SR)}$ in terms of G in SI units.

INDEX

Q

R

S